HISTORY OF TECHNOLOGY

MICHAEL FARADAY
THE INSTITVTION OF
ELECTRICAL ENGINEERS

History of Technology

Volume Thirteen, 1991

Edited by
Graham Hollister-Short and
Frank A.J.L. James

MANSELL

LONDON AND NEW YORK

First published 1991 by
Mansell Publishing Limited, *A Cassell imprint*
Villiers House, 41/47 Strand, London WC2N 5JE, England
387 Park Avenue South, New York, NY 10016-8810, USA

British Library Cataloguing in Publication Data

History of technology.—13th volume (1991)
1. Technology—History—Periodicals
609 T15

ISBN 0-7201-2114-0
ISSN 0307-5451

Library of Congress Catalog
Card Number: 76-648107

Frontispiece: Michael Faraday, from an IEE bookplate, 1908, by George W. Eve

Printed and bound in Great Britain
by Biddles Ltd, Guildford and King's Lynn

HISTORY OF TECHNOLOGY

Contents

Editorial

In this volume we break new ground in two distinct ways. For the first time our contributors will be found addressing a single thematic issue. At the same time the theme in question, the history of electrical technology, is virtually a new one for our pages. Previously only three contributions, those by Keith Dawson in Volume 1, D.G. Tucker in Volume 7 and D. de Cogan in Volume 10, barely 3 per cent that is to say, of the papers in our twelve earlier volumes have been solely concerned with aspects of electrical technology.

For a subject of such importance to make so poor a showing as this is, in our opinion, more than a little surprising. When one considers the life (and mind) transforming function which electrical devices exercise on the existence of virtually everyone now alive, not to mention the protean role played by electricity in the development of so many technologies at the cutting edge of twentieth-century engineering, it is obvious that the history of electrical technology has (for whatever reason) been strangely neglected. If the neglect in question had affected only this journal then that would be merely one more of our parochial problems. In fact the subject has been only slightly better served in the other major journals devoted to the history of technology. It is not without interest that it has not been selected as a theme at any of the eighteen ICOHTEC conferences so far held. With the present volume we begin at least to redress the imbalance.

Neglect of the subject is surprising for another reason, since it is undoubtedly a fruitful area for research into the important historiographical question concerning the nature of the relationship between science and technology. The rise of electrical technology was taking place at much the time when it was becoming obvious that whatever might have been the connection between science and technology in earlier times, 'know why' and 'know how' were now entering, in the last quarter of the nineteenth century, some qualitatively different type (but what type exactly?) of relationship. Electrical technology was not of course simply coeval with this change but was part of the change itself. To make it the focus of one's study is therefore, in heuristic terms, to put oneself on the road to understanding the epistemological complexities wrapped up in what might be called provisionally (and loosely) 'scientific technology'. The relationship of science and technology was certainly not a straightforward case of simple causation with scientific work on electricity leading to the development of electrical engineering. Scientific knowledge was a necessary, but not a sufficient, condition for technological advance. In electrical engineering scientific method had to be applied to engineering practice. Knowledge of materials was of crucial importance. So too was the 'hands-on' knowledge possessed only by skilled craftsmen deploying craft skills derived from other already existing technologies.

The occasion for this thematically unified issue is the bicentenary of the birth of Michael Faraday this year. Celebration will take many forms. We thought it would be appropriate to present our contribution in the form of a volume honouring a man generally credited with the discovery of the principles of the electric motor, the transformer and the dynamo, one who is certainly looked upon by professional electrical engineers as the founder of their discipline. This also explains our choice of frontispiece. Lest we be suspected of hagiographical inclinations we must say at once that it was only the work of many later individuals that cumulatively conferred on Faraday's work the importance that we see it has today. It is true, nevertheless, that, as Charles Singer once said, 'it is the first steps that count'.

For this reason the first three papers deal with the contemporary impact of Faraday's work on telegraphy (the first electrical technology) and on the development of the electric motor. The next two papers (by Graeme Gooday and the late Bill Reader) deal with institutionalization of electrical engineering in Britain at the end of the nineteenth century. It was this process that facilitated the entry of many more men (and eventually women) into the new profession of electrical engineering. The work of two of these men, Jenkin (the man of practice) and Heaviside (the man of theory) are discussed in the two papers following.

One feature of the history of electrical engineering is of special interest. Although the new technology was largely initiated in the traditional area of technical innovation, namely Western Europe, its subsequent diffusion did not follow the pattern that had characterized the movement of innovations in the earlier decades of the industrial revolution, that is to say westwards to the United States. American developments in electrical enginnering were fundamental to its future growth and flowed back into Europe, along with, it may be noted, numbers of other items, each returning, as one might say, with interest. The complex international relations which existed in electrical engineering (as in the technology of the day generally) form the focus of the four concluding papers. The American contribution was, as Brian Bowers indicates, in many senses of the word a central one: the first central generating plant in London was American-built and was built before any such plant was operating in the United States itself. One finds historians and commentators contemporary with this period speaking of the 'three great technological languages: English, French and German' but correct observation though this was, it concealed another, larger, truth: by 1900 the technological world was bipolar and Western Europe was not as alone as the test of language might suggest. With the coming of electrical engineering the pre-eminence of North America in that field was a clear enough signal that a second great focus of activity had come into being.

G.H.-S F.A.J.L.J

The Contributors

Brian Bowers is Senior Curator, Electrical Engineering, in the Science Museum. His major research interests, on which he has written and lectured, include the development and application of electricity supply, lighting, the work of Wheatstone and Faraday and Edison's activity in Britain. *Address*: Science Museum, Exhibition Road, South Kensington, London SW7 2DD, England.

R.W. Burns recently retired as head of the Department of Electrical Engineering at Trent Polytechnic. He has written widely on the history of television, including *British Television, the Formative Years* (Stevenage, 1986). *Address*: 29 Dunster Road, West Bridgford, Nottingham NG2 6JE, England.

W. Bernard Carlson teaches history of technology in the Humanities Division, School of Engineering and Applied Science, University of Virginia. He is author of *Innovation as a Social Process: Elihu Thomson and the Rise of General Electric, 1870–1900* (Cambridge, forthcoming), in which he develops the concept of craft knowledge. With Michael E. Gorman he is currently studying invention as a cognitive process by comparing the telephone inventions of Alexander Graham Bell, Thomas Edison and Elisha Gray. *Address*: Humanities Division, School of Engineering and Applied Science, University of Virginia, Thornton Hall, Charlottesville, VA 22903, USA.

Jonathan Coopersmith is an assistant professor of history at Texas A & M University. His main interest is in Russian and Soviet technology. *Address*: Department of History, Texas A & M University, College Station, Texas 77843-4236, USA.

Brian Gee is a freelance author with a particular interest in scientific practitioners and engineers in the 1830s. *Address*: 18 Barton Close, Landrake, Saltash, Cornwall PL12 5BA, England.

Graeme Gooday is a British Academy Post-Doctoral Fellow at the University of Kent. His current research is in the social history of nineteenth-century laboratories, particularly in relation to the sociology of experimental education in British physics and electrical engineering. *Address*: History of Science Unit, Physics Laboratory, University of Kent, Canterbury, Kent CT2 7NR, England.

Colin Hempstead is Senior Lecturer in the History of Science and Technology at Teesside Polytechnic. His research interests are broadly concerned with the history of electricity and electrical engineering. The present paper stems from these studies and from the fact that two of his granddaughters are great-great-great-granddaughters of Fleeming Jenkin. *Address*: Department of Humanities, Teesside Polytechnic, Borough Road, Middlesbrough, Cleveland TS1 3BA, England.

Bruce Hunt teaches history of science at the University of Texas at Austin. He recently completed *The Maxwellians* (Cornell, forthcoming), a study of British electrical theory in the 1880s and 1890s, and is now continuing his examination of the interaction between cable telegraphy and electrical science in the mid-nineteenth century. *Address*: Department of History, University of Texas, Austin, TX 78712-1163, USA.

A.C. Lynch was at the Post Office Research Station, Dollis Hill, London, from 1936 to 1976 and is now an Honorary Research Fellow at University College London and a consultant to the National Physical Laboratory. Since leaving full-time research he has studied and written on the history of electrical measurements in the late nineteenth century. *Address*: 8 Heath Drive, Potters Bar, Hertfordshire EN6 1EH, England.

Iwan R. Morus is Wellcome Fellow at the Wellcome Unit for the History of Medicine at Cambridge University. He is interested in electricity and life, in the science and technology of the nineteenth century. *Address*: Wellcome Unit for the History of Medicine, Department of History and Philosophy of Science, University of Cambridge, Free School Lane, Cambridge CB2 3RH, England.

W.J. Reader was an historian by training. He wrote the two-volume history of Imperial Chemical Industries and was the author of the official history of the Institution of Electrical Engineers. He died in June 1990.

Notes for Contributors

Contributions are welcome and should be sent to the editors. They are considered on the understanding that they are previously unpublished in English and are not on offer to another journal. Papers in French and German will be considered for publication, but an English summary will be required. The editors will also consider publishing English translations of papers already published in languages other than English. Three copies should be submitted, typed in double spacing (including quotations and notes) with a margin on A4 or American Quarto paper. Include an abstract of 150–200 words and two or three sentences for 'Notes on Contributors'.

It would be appreciated if normal printers' instructions could be used. For example, words to be set in italics should be underlined and *not* put in italics. Authors who have passages originally in Cyrillic or oriental scripts should indicate the system of transliteration they have used. Quotations when long should be inset without quotation marks; when short, in single quotation marks. Spelling should follow the *Oxford English Dictionary*, and arrangement H. Hart, *Rules for Compositors* (Oxford, many editions). Be clear and consistent.

All papers should be rigorously documented, with references to primary and secondary sources typed separately from the text in double spacing and *numbered consecutively*. Cite as follows for books:

1. David Gooding, *Experiment and the Making of Meaning: Human Agency in Scientific Observation and Experiment*, (Dordrecht, 1990), 54–5.

Subsequent references may be written:

3. Gooding, *op. cit.* (1), 43.

Only name the publisher for good reason. For theses, cite University Microfilm order number or at least Dissertations Abstract number. Standard works like DNB, DBB may be thus cited.

And as follows for articles:

13. Andrew Nahum, 'The Rotary Aero Engine', *Hist. Tech.*, 1986, 9: 125–66, p. 139.

Line drawings should be drawn boldly in black ink on stout white paper, feint-ruled paper or tracing paper. Photographs should be glossy prints of good contrast and well matched for tonal range. The place of an illustration should be indicated in the margin of the text where it should also be keyed in. Each illustration must be numbered and have a caption. Xerox copies may be sent when the article is first submitted for consideration.

Michael Faraday, Cable Telegraphy and the Rise of Field Theory

BRUCE J. HUNT

Michael Faraday's field theory of electricity and magnetism is now regarded as one of the great successes of nineteenth-century physics, but it attracted little real support before the mid-1850s. Indeed, the basic idea that electric and magnetic phenomena resulted from stresses and strains distributed through space had been proposed by others long before Faraday but had repeatedly lost out to the more precise and less hypothetical action-at-a-distance theories.[1] When Faraday began to elaborate his field approach in the 1830s and 1840s, most theorists dismissed his ideas about lines of force and 'fields' as little more than a mental crutch suitable for one like Faraday who was unable to handle a proper mathematical theory. From the mid-1850s, however, Faraday's approach began to gain wider support, and by the time James Clerk Maxwell's *Treatise on Electricity and Magnetism* appeared in 1873, field theory had become virtually the new orthodoxy—but only in Britain. Although German physicists read Faraday's papers and admired his experimental discoveries, they remained sceptical of his theoretical ideas and did not take up the field approach in any numbers until the late 1880s and early 1890s, in the wake of Heinrich Hertz's experiments on electromagnetic waves.[2]

Why did the British and German responses to Faraday's ideas differ so widely in the third quarter of the nineteenth century? What created a specifically British market for the rather odd cluster of ideas known as field theory, and why did this market emerge only in the mid-1850s and 1860s? I will argue that at least part of the answer lies in differences in the British and German technological environments in this period, in particular in the needs and opportunities presented by the emerging submarine cable industry. Cable telegraphy was, from its earliest days, a virtual British monopoly, and it exerted a major influence on the direction and content of British electrical research throughout the second half of the nineteenth century. Faraday made a point of relating his theories to the new phenomena encountered on cables in the 1850s, and one of my aims will be to show how his ideas were taken up, put to use and extended by those active in cable telegraphy. I shall be concerned not with the origin of Faraday's field ideas, but with their reception and spread, and with the context in which they were put to use and judged to have advantages over other approaches.[3] This, as we shall see, was largely in the treatment of 'retardation' and other propagation phenomena encountered on underground and underwater telegraph lines.

FARADAY AND RETARDATION

Although Faraday had close ties with Charles Wheatstone and others involved in the practical applications of electricity, his own role in early telegraphy was limited and mostly indirect. One of his main contributions came in 1848, when he helped to publicize the insulating properties of gutta-percha, a rubber-like vegetable gum then newly introduced from Malaya.[4] The Gutta-Percha Company of London soon began using it to cover telegraph wires to be laid underground or underwater, and in 1851 the first successful submarine cable—four copper wires insulated with gutta-percha and protected by an outer rope of iron wires—was laid across the English Channel by J.W. Brett and the Submarine Telegraph Company. It proved extremely profitable and was soon followed by other relatively short cables in northern Europe and the Mediterranean, almost all laid by British companies. There was even talk of laying cables to India, Australia and America. As the leading industrial, commercial and imperial power of the day, Britain had both the means and the incentive to exploit this important new technology, and after pioneering cable telegraphy in the 1850s and 1860s, the British continued to dominate it well into the twentieth century.

Even before the first cables were laid, the growth of the overland telegraph network had begun to attract the attention of the Astronomer Royal, George Biddell Airy, who wanted to use the wires to distribute accurate time signals and to connect distant observatories for accurate longitude determinations. From about 1849, he was in close touch with a number of telegraph engineers, including C.V. Walker of the Southeastern Railway telegraph department, Edwin and Latimer Clark of the Electric Telegraph Company and C.S. Wollaston of the Submarine Telegraph Company, and in 1852–3 he had wires installed connecting Greenwich Observatory to their networks.[5] Early in 1853, Airy made telegraphic determinations of the longitudes of the observatories at Cambridge and Edinburgh, and he began to plan similar determinations via cables to Paris, Brussels and Dublin, and via the Electric Telegraph Company's new underground lines to Liverpool and Manchester.

Latimer Clark took a particular interest in Airy's work, lending him batteries and equipment and arranging his use of the Electric Telegraph Company lines. Clark had also been making his own studies of telegraphic phenomena, and in October 1853 he wrote to tell Airy that:

> We have lately observed a great and variable retardation of the Electrical Current when sent through long lengths of underground wire, which would much interfere with your use of Subterranean or Submarine wires for purpose of determining longitudes, and I think it right therefore to call your attention to it. Professor Faraday will

> attend at Lothbury on Saturday evening next at 5 o'clock to repeat some experiments in which this retardation will be well exhibited, and we should be happy for you to be present at the same time if you consider it desirable.[6]

Exactly when Clark first invited Faraday to see his experiments is unknown, but his reasons for doing so were straightforward: besides wishing to bring an interesting new phenomenon to Faraday's attention, Clark wanted advice from Britain's leading electrical expert on how to overcome what threatened to become a serious practical problem. Clark had found that a short sharp pulse sent in at one end of a long cable or underground line emerged at the other badly blurred and stretched out, and only after an appreciable delay. If no way could be found to counteract this 'retardation' it would limit the rate at which signals could be sent effectively through a cable or underground line—if one tried to send too rapidly, the result at the far end was an indecipherable blur—and so limit the amount of business the line could handle in a day. Moreover, the retardation got worse on longer cables and underground lines, suggesting that signalling through a really long cable, such as one across the Atlantic, might be too slow to be profitable.

Clark had first observed retardation about a year earlier during tests with Samuel Statham, head of the Gutta-Percha Company Works at Wharf Road near the Thames. Clark published a short account of his findings in the *Chemical Record and Drug Price Current* in March 1852, but it attracted no notice.[7] He now used the interest stirred up by Airy's longitude studies as an opportunity to air his discovery more widely in the scientific community. He arranged for Faraday to be shown demonstrations of the chief phenomena of retardation, first at the Gutta-Percha Company Works on 4 October 1853 and then at the Electric Telegraph Company office at Lothbury in the City later that day and on 15 October.[8] These were Clark's and Statham's experiments; Faraday was merely an observer. A 110-mile length of insulated wire being prepared for a Mediterranean cable was slung in coils over the side of a canal barge at the Gutta-Percha Company Works, and Faraday was shown that when this 'water wire' was connected to a battery, it took up a very large charge, much like a huge Leyden jar. This gave a direct demonstration of the identity of static and dynamic electricity: the galvanic current flowed into the wire until it had established a substantial static charge, and the subsequent gradual discharge of the wire produced all the same effects, both magnetic and electrochemical, as an ordinary battery current. The time taken to charge and discharge the wire was the most notable feature of the phenomenon, and it showed itself only with the 'water wire'; 100 miles of insulated wire kept in air showed no detectable effect. Some of the effects were quite striking: for instance, Faraday was shown how a current could be applied to one end of the long water wire, with the far end put to earth, and then nearly all drawn back out at the near end by putting it quickly to earth through a galvanometer. The phenomena were clearly of a scale and type unattainable in an ordinary laboratory.

Clark next took Faraday to the Electric Telegraph Company offices and showed him similar phenomena in long underground wires running to Liverpool and back. More elaborate demonstrations were given there on 15 October, with Airy and several telegraph men also in attendance.[9] This time the eight underground lines to Manchester were all connected together, giving a total of nearly 1,600 miles of wire, with galvanometers connected on the return loops in London to show the passage of the current along the way. The retardation of the signals was obvious: a current applied in London caused the galvanometers to deflect in turn rather than all at once. The blurring of the pulses could also be seen: what had started as a sharp deflection of the first needle became a more and more gradual swing of the later ones. This was even more striking when the galvanometers were replaced by Bain chemical telegraphs, which made marks on a moving paper tape. These gave a permanent and measurable record of the retardation and blurring, and it was clear, as Faraday remarked in his notes, that 'the wave of power' passed more 'sluggishly' along the underground wires than along a parallel overhead wire, on which the signals returned sharply and almost instantaneously.[10] These findings were reinforced in November and December during Airy's Brussels longitude tests, in which the current was found to take about 0.1 seconds to pass through the underground and submarine wires connecting Greenwich to Brussels—far more than the barely detectable retardation found on the much longer Greenwich–Edinburgh circuit, most of which ran on overhead wires. Airy drew attention to this retardation in an article in the *Athenaeum* in January 1854, ascribing it to 'an ill-understood effect of induction'.[11]

After being shown a few further experiments at the Gutta-Percha Company Works early in January, Faraday gave an account of what he had seen in a Friday evening lecture at the Royal Institution on 20 January 1854. The *Philosophical Magazine* published what it called a 'verbal copy of an abstract' of this 'important paper' in March, and in later years it was often cited as the foundation of the theory of retardation phenomena.[12] Faraday began by staking a claim on behalf of himself and other scientists: 'when the discoveries of philosophers and their results are put into practice,' he said, 'new facts and new results are daily elicited', as recent work in telegraphy had strikingly demonstrated.[13] After thanking Statham and Clark for their help, he declared that their experiments provided 'remarkable illustrations of some fundamental principles of electricity, and strong confirmation of the truthfulness of the view which I put forth sixteen years ago [in 1838]' on the relations between 'induction, conduction, and insulation.'[14] He proceeded to describe the retardation phenomena he had seen and to explain them in terms of his own theories of electric action. The action-at-a-distance theory could account for retardation only by making additional hypotheses, and gave no obvious reason why the electric fluid should move so differently along an insulated wire running under water than it did along one that was strung from poles. But Faraday regarded the insulating material *around* the wire as the real seat of electromagnetic phenomena, and he had no trouble explaining why pulses sent along a cable should be retarded. The conduction of a current was always preceded, he said, by the induction of a state of strain in the surrounding dielectric and the consequent storage of a certain amount of charge. When the inductive capacity was large, as in a long cable, it took an appreciable time to induce this strain, and the rise of the current was thus somewhat delayed and smoothed out—as was the subsequent fall of the current when the circuit was broken. Conversely, when the inductive capacity was small, as in ordinary overhead lines, the strain was induced almost instantaneously and no retardation was observed.

All these phenomena depended crucially on the nature and arrangement of the surrounding dielectric—a thin layer of gutta-percha sandwiched between two conductors in the case of a cable, fifteen or twenty feet of air in the case of an overhead wire. Cable telegraphy showed in a striking and palpable way the *propagation* of electrical actions in time that Faraday had previously been able to assert only theoretically; moreover, it showed that the speed of this propagation depended as much on the surrounding dielectric as on the conductor itself. Experience with cable propagation tended to shift attention away from the conductor and towards the surrounding medium. In particular, it brought out the importance of the 'specific inductive capacity' of different dielectrics, by which electrical phenomena depended not just on the arrangement and state of the conductors, but on the nature of the intervening medium as well. The idea that dielectrics played such an active role in electrical phenomena had been widely dismissed when Faraday first put it forward in 1842, but the advent of cable telegraphy a decade later made specific inductive capacity the focus of great practical and scientific interest. As Latimer Clark told an official investigation of cable construction in 1860, 'At the date of Faraday's interesting researches [on specific inductive capacity], it could little be foreseen that such an obscure phenomenon should be destined to become one day, as it has now, a consideration of high national importance, and one which has a direct and most important bearing on the commercial value of all submarine telegraphs.'[15]

PRUSSIAN INTERLUDE

A brief digression will illustrate how cable telegraphy helped to focus attention on the role of the dielectric, and will reinforce the point that retardation was a specifically British concern in the mid-1850s and 1860s. Latimer Clark had in fact not been the first to observe the inductive charging of underground wires. In 1848, Werner Siemens, then a lieutenant in the Prussian army, was ordered to lay a line as quickly as possible from Berlin to Frankfurt, where the new German national congress was then meeting. (The Prussian authorities wanted to monitor the activities of the congress and ordered that the line be laid underground to protect it from damage by the 'turbulent' population.) Several hundred miles of gutta-percha covered wire were hastily manufactured and laid two or three feet deep along the roads and railway lines.[16] Siemens soon noticed peculiar charging effects on his wires and concluded, as Clark and Faraday were to do a few years later, that they were acting as enormous Leyden jars—what he called 'jar wires'. He mentioned this observation in an account of his telegraphic work at the French Academy of Sciences in 1850, but it attracted little notice at the time.[17]

Siemens's experience with the 'jar wires' led him to adopt a Faraday-like view of electrical action in which inductive effects were traced to 'molecular polarization' of the surrounding medium rather than to direct action at a distance. He was too busy with his business affairs to write up a full account of his new views in the early 1850s, but in 1857 he published a long paper in the *Annalen der Physik*, 'Ueber die electrostatische Induction und die Verzögerung des Stroms in Flaschendrähten' ('On electrostatic induction and retardation of the current in cores'), that was devoted almost entirely to an elaboration of the view, partly derived from Faraday and partly developed by Siemens himself, that electrical actions were due to the polarization of the surrounding dielectric. Indeed, Siemens concluded this paper by stating that 'it is very likely that the seat of the electricity is removed from the conductors to the non-conductors surrounding them, and may be defined as an electrical polarization of the molecules of the latter'.[18] This shift of focus from the conductor to the dielectric was central to the field approach that was then gaining ground in Britain, but it remained unusual in Germany.[19]

By the time Siemens's paper appeared, the underground lines that had led him to take this view no longer existed. The insulation on the Berlin–Frankfurt line and on the rest of the Prussian underground network had failed within a year or two of its being laid, as the use of poorly treated materials, improper maintenance and the depredations of foraging rats combined to destroy the gutta-percha covering. Siemens (who had resigned from the army) was dismissed by the civilian authorities and the underground lines were pulled up and replaced by overhead wires in the early 1850s.[20] With no underground lines (and no submarine cables), German scientists and engineers had neither the need nor the opportunity to confront retardation phenomena after the early 1850s, and so paid little attention to the more revealing aspects of telegraphic propagation. Significantly, Siemens transferred his cable business to his brother William, who was based in London.[21]

WILLIAM THOMSON'S CABLE THEORY

I shall now return to events in Britain. It is well known that William Thomson (later Lord Kelvin) was one of the first physicists to take up Faraday's ideas and cast them into mathematical form, and also that he was a leading figure in the British cable industry. My aim here will be to show some of the connections between these two aspects of Thomson's work. Thomson's involvement with cables began around 1853–4 and continued throughout the rest of his life, providing the foundation of his personal fortune and much of his public reputation. Cable telegraphy also had a deep effect on his choice of problems to study and on the way he conceived of electromagnetic phenomena, providing, as Crosbie Smith and Norton Wise have recently emphasized, a guiding thread through his electrical career.[22]

Thomson was probably first exposed to the cable business through Lewis Gordon, his colleague as professor of engineering at Glasgow University and a partner in R.S. Newall's wire rope and cable-making firm. (Several of the early submarine cable firms started as wire rope manufacturers; Newall was the most energetic and adventurous of these, though a series of failures in the late 1850s and 1860s eventually led him to withdraw from the cable business.) By December 1853, Thomson was sufficiently interested in cable propagation to ask (through Edward Sabine) for information from Airy about the circuits used in the Brussels longitude tests and how their arrangement had affected the speed of the current.[23]

Faraday's January 1854 lecture also caught Thomson's interest; indeed, it may well have been what prompted him to dust off two papers on the mathematical theory of electricity he had written in the 1840s and send them to the *Philosophical Magazine* for republication.[24] The first of these, published anonymously in the little-read *Cambridge Mathematical Journal* in 1842, concerned the analogy between the flow of heat and the distribution of electric potential. Thomson, then a very young undergraduate, showed that Fourier's equations governed both sets of phenomena, so that despite their evident physical differences, problems both of heat and of electricity could be expressed interchangeably in terms either of action at a distance or of contiguous action.[25] The second paper, written while Thomson was in Paris and published in 1845 (in what had become the *Cambridge and Dublin Mathematical Journal*) used this method to show that experiments by Faraday and William Snow Harris, and Faraday's field treatment of electrical forces, did not conflict with Coulomb's inverse square law of attraction and repulsion. Although often later cited as a step towards Thomson's adoption of field theory, the paper is perhaps better read as an attempt to defend the validity of Coulomb's phenomenological law while leaving the question of the physical nature of electrical forces open.[26]

Whatever role Faraday's lecture played in prompting Thomson to republish his 1842 and 1845 papers, it certainly lay behind a third paper, intended as a supplement to the first two, that he sent off to the *Philosophical Magazine* later in 1854. This concerned the calculation of the capacity of a Leyden jar or 'a telegraph wire insulated in the axis of a cylindrical conducting sheath'—that is, he said, 'the copper wires in gutta-percha tubes under water, with which Faraday has recently performed such remarkable experiments.'[27] (Like many others, Thomson persisted in crediting these experiments to Faraday, but they were really Clark's and Statham's, as Faraday—and Clark—always tried to make clear.) Thomson completed and sent this third paper to the *Philosophical Magazine* in June 1854, but it was not published until about a year later. The reason for the long delay is unclear, but it may have been related to Thomson's efforts to secure a patent on a new type of cable conductor; concern on this score certainly led him to hold back the release of related work later in 1854.[28]

Cable questions were thus already much on Thomson's mind when he set off for the Liverpool meeting of the British Association in September 1854. Several cable engineers and promoters—including J.W. Brett, C.F. Varley, Frederick Bakewell, E.B. Bright and J.B. Lindsay—gave papers there on various projects for cables to India and America and on phenomena they had observed on underground and submarine lines. Varley and Bright made particular reference to retardation effects and possible means of overcoming them.[29] It is unclear what part Thomson took in the discussion of these papers, but an incident at the very end of the meeting prompted him to produce his most important contribution to cable theory.

Airy was again involved, though at a couple of removes. He had written the previous year to his Irish counterpart, William Rowan Hamilton, about the possibility of connecting the Greenwich and Dunsink observatories by telegraph—after several failed attempts, a cable had been laid from Scotland to northern Ireland in May 1853.[30] Hamilton's son Archibald was interested in electricity and had done some telegraphic experiments at Dunsink in which he used ground and water conduction to signal without connecting wires. He came along to Liverpool, and at the end of the meeting he and his father asked Thomson whether a similar method might be made to work across the Irish Sea. But Thomson was obliged, as he later said, 'to run away to get to a steamer by which I was bound to leave for Glasgow', and he passed the Hamiltons and their question to his friend G.G. Stokes, who took up the subject with his customary thoroughness.[31]

Stokes proceeded to work out the problem in great detail and soon sent young Hamilton a long letter (now lost) showing mathematically that his idea would not work.[32] Stokes had not previously given much attention to electrical problems, but this exercise sparked his interest in the theory of ordinary cable signalling and in October 1854 he wrote to ask Thomson about it. In reply, he received two long and important letters in which Thomson laid out what was to become the accepted theory of telegraphic propagation. By translating Fourier's equations for the propagation of heat into electrical terms and combining them with Faraday's ideas about specific inductive capacity, Thomson was able to derive a simple equation governing the rise of current and voltage at the far end of a submarine cable: $ck\, \mathrm{d}v/\mathrm{d}t = \mathrm{d}^2 v/\mathrm{d}x^2$, where v is the potential at a point and c and k are the capacitance and resistance per unit length. Oliver Heaviside later called Thomson's theory 'the first step towards getting out of the wire into the dielectric' in the treatment of cable propagation—a step Heaviside was himself to follow up in the 1880s with his theory of transmission wires as guides for electromagnetic waves that slipped along outside them.[33]

Thomson's main finding was that the retardation was proportional to both the capacitance and the resistance of the cable; since, for a given gauge, these were both proportional to the length of the cable, the retardation increased with the square of its length—the so-called 'law of squares'. Short cables could thus be worked much more quickly than long ones, and since the resistance decreased with the thickness of the wire and the capacitance with thickness of the gutta-percha (relative to the copper), thick cables were better than thin ones. Thomson wrote to Airy in January 1855, explaining his theory and asking about the dimensions of the wires used in the Brussels longitude tests. Given the retardation of 0.1 seconds found in those tests, the law of squares implied that an Atlantic cable of the same gauge (and nearly ten times the length) would suffer a retardation of about ten seconds—enough, Thomson said, to make it 'almost useless'.[34] A thicker cable would suffer proportionally less retardation, and Thomson hoped to use the information on the Brussels wires to determine whether an Atlantic cable able to carry a reasonable number of messages per day would have to be so thick as to be mechanically and financially unwieldy. He also hoped to use the data to make a rough estimate of the ratio of electromagnetic to electrostatic units, an important quantity whose value was then completely unknown.

Airy forwarded Thomson's request to C.S. Wollaston of the Submarine Telegraph Company, adding that the information would aid Thomson in 'theoretical investigations on the transmission of galvanic currents which appear likely to lead to results of great practical importance'.[35] Wollaston replied with full particulars, saying he was happy to help and expressing the hope that Thomson would be able to shed some light on the question of telegraphic transmission, of which 'little is at present known either theoretically or practically'.[36] After receiving Wollaston's data, Thomson sent Airy a long letter detailing his latest thoughts on retardation and the ratio of units, which he was able to estimate within a factor of about two. He apologized for writing at such length about such things to Airy, but it was, he said, 'a subject in which I take much interest myself'.[37]

Thomson was 'anxious to get something brought out' on all this, and shortly after writing to Airy he asked Stokes (then newly elected Secretary of the Royal Society) if he could arrange for extracts from his letters to be published in the *Proceedings* of the Society.[38] A paper consisting largely of Thomson's letters to Stokes of 28 and 30 October 1854 (and part of one of Stokes's replies) duly appeared in May 1855 as 'On the theory of the electric telegraph'. It was reprinted early the next year in the *Philosophical Magazine* and was later widely read and cited.[39] Thomson repeated his conclusions at the 1855 British Association meeting and expressed even more doubts about the practicability of signalling at a reasonable rate through long cables. Citing the 350-mile line Newall had laid from Varna to Balaklava that May to provide communications during the Crimean War, Thomson said that an Atlantic cable (six times as long) able to work as quickly—and the Black Sea cable could apparently handle only about five words per minute—would have to be six times as thick and so, in copper and gutta-percha alone, cost 216 times as much.[40]

These estimates posed a threat to the plans—already being vigorously promoted by the American Cyrus Field—to lay a cable across the Atlantic. If Thomson's theory were correct, Field's project would be a bad investment; a cable that was thin and so relatively cheap to make and lay would not be able to carry enough traffic to be worthwhile, while one that was thick and so relatively fast would cost too much and be too bulky to be built and laid in the first place. The issue was put in exactly these terms by E.O. Wildman Whitehouse, a Brighton surgeon who had taken up electrical experimentation with J.W. Brett in the early 1850s and become involved in the plans for an Atlantic cable. At the 1856 meeting of the British Association at Cheltenham, Whitehouse denounced Thomson's 'law of squares' as a 'fiction of the schools' and said that his own experiments showed that a cable of ordinary thickness would be able to carry at least several words per minute across the Atlantic.[41] Indeed, Whitehouse claimed that his experiments showed that increasing the cross-section of a wire *increased* the retardation along it, in direct contradiction to Thomson's theory.

Thomson was not at the Cheltenham meeting, but he wrote to the *Athenaeum* a few weeks later to reiterate his confidence in his theory and to say that Whitehouse had misinterpreted it.[42] The peculiarities of the sending and receiving apparatus had to be taken into account when calculating the signalling rate to be expected for different lengths, he said, and when this was done, Whitehouse's experiments turned out to *support* the law of squares. But Thomson had to admit that he had been wrong to say that an Atlantic cable would have to be six times as thick as the Black Sea cable—and of course it was this, not arcane points of theory, that Whitehouse had really objected to. Thomson had assumed that the problems with rapid signalling on the Black Sea cable had been due to retardation, but it now appeared that they were caused mainly by the use of inappropriate sending and receiving instruments. New measurements of the conductivity of copper by Wilhelm Weber in Germany, and new estimates of the specific inductive capacity of gutta-percha, now enabled Thomson to calculate the amount of retardation on a cable from first principles, and while he found that it would still be a problem on long cables, he concluded that it would not be as serious as he had originally thought.[43]

By November 1856, Thomson and Whitehouse were able to agree that retardation would not make an Atlantic cable impracticable. This was fortunate, since Whitehouse had just been appointed the official 'electrician' of Field's newly formed Atlantic Telegraph Company and Thomson was soon to be named to its Board of Directors. But neither Thomson and Whitehouse nor the rest of the electrical community had yet reached a consensus on the laws governing retardation or on the best design for the proposed cable. Should the cable be *thin*, in accordance with Whitehouse's ideas about reducing retardation and to save money (and, according to one report, to look nice when showing samples to investors)?[44] Or should it be relatively *thick*, as Thomson thought best in order to lessen the resistance and so the retardation, and as some of the company's engineers preferred for mechanical strength?[45] All of this was complicated by Cyrus Field's great haste to get the project going, mainly for financial reasons, and by the need for all of those associated with the Company to keep up an air of confidence in order not to frighten investors away from what was, after all, an extremely ambitious and risky project.

These issues surfaced publicly—and Faraday briefly re-entered the story—on 13 January 1857, when F.R. Window, a telegraph engineer unconnected with the Atlantic project, gave a paper 'On submarine electric telegraphs' at the Institution of Civil Engineers in London. After tracing the history of the first cables, Window turned to the problem of retardation, describing Clark's experiments at considerable length and explaining Faraday's theory of 'lateral induction'. 'Although a pretty and valuable experiment to the philosopher', Window said, retardation was 'a great obstruction to the working of telegraphs'; indeed, he calculated that the proposed Atlantic cable would be able to transmit no more than about one word per minute.[46]

The backers of the Atlantic cable saw these criticisms as a renewed threat and moved quickly to answer them—indeed, Window felt constrained to open the discussion (which filled the whole of the next two meetings) by saying that he had not meant 'to disparage the attempt to convey telegraphic messages to America, nor did [he] mean to say, that the speed would be so slow as to be a bar to profit.'[47] Charles Bright, newly hired as the chief engineer of the Atlantic Telegraph Company, rose to declare that the effects of retardation had been exaggerated and that Whitehouse's latest experiments had shown that ten to twelve words per minute could be sent through 2,000 miles of underground line, and presumably through the same length of submarine cable as well. He added that 'Professor Thomson, of Glasgow, was at one period a great opponent of the theory of the practicability of working through long distances, but he has seen reasons to change his opinion on the subject, and is now one of its warmest supporters.' Thomson himself wrote a letter, quoted by Bright at the meeting, saying that his earlier objections had been answered and that he now believed that the proposed Atlantic cable would be able to handle at least three words per minute and perhaps considerably more once Whitehouse's transmission instruments were perfected.[48] Whitehouse also spoke, as did Latimer Clark, C.F. Varley and several other prominent telegraph engineers. William Siemens was virtually the only speaker to echo Window's doubts about transatlantic signalling rates.

Faraday attended the second discussion session on 27 January (apparently a somewhat unusual step for him), and while he said that 'in stating his view of the cause' of retardation he 'would endeavour to avoid referring to his own theoretical views, because the facts stood apart from all theory', his theories in fact underlay all of his remarks. As in 1854, he compared a cable to a Leyden jar and said that 'lateral induction' across the gutta-percha temporarily diverted the 'propulsive force' of the current and so retarded its propagation. 'The larger the jar, or the larger the wire, the more electricity was required to charge it,' he said, 'and the greater was the retardation of the electric impulse, which should be occupied in sending the charge forward.'[49] This was the key point—Faraday seemed to be saying that the low capacitance of a thin cable would enable it to transmit signals *faster* than a thick cable. Certainly his remark was later cited to this effect by Field and others, who pointed to it as an authoritative endorsement of the thin cable they already preferred on grounds of financial expediency.[50]

In fact Faraday had made a serious error. In 1854 he had said that retardation arose from the *combined* effects of resistance and capacitance, as Thomson was to show in more detail a little later. In now implying that a thin cable would produce less retardation than a thick one, Faraday was focusing solely on the lower capacitance of a thin wire (when covered with the same thickness of gutta-percha) and ignoring its higher resistance. The increased resistance of a thin wire would in fact swamp the slight decrease in its capacitance and, as Thomson had shown, its retardation would be much greater than that on a thick cable.

Faraday's other remarks at the meeting were cautious but supportive of the Atlantic cable project. He thought it quite possible that use of 'curbing' (in which a sharp positive pulse was followed by a negative one to help 'clear the charge') would increase the signalling rate, but he was hesitant to commit himself without having done experiments. Even without such improvements, he thought signals would take no more than about two seconds to pass through the proposed Atlantic cable, and this seemed sufficient for commercial purposes. In any case, he thought that 'the experiment of a transatlantic line is one which two such countries as England and America should glory in carrying out'.[51]

FAILURE AND RATIONALIZATION

The subsequent history of the Atlantic cable has been told many times, and I will sketch only a few relevant points.[52] Field and the Atlantic Telegraph Company were happy to have Faraday's apparent endorsement of a relatively thin cable, since they had already ordered one of substantially that design. Thomson went along, apparently believing it to be the best cable that could be financed at the time, and Newall and Glass, Elliot began manufacturing it in January 1857.[53] (The order was split partly to speed production and partly to keep both of the main cable suppliers happy. The work was so badly coordinated that Newall used a right-handed 'lay' for the outer protective wires, while Glass, Elliot used a left-handed one, requiring an elaborate splice in the middle.) After an abortive attempt to lay it in August 1857, the cable was stored (rather badly) on the docks at Plymouth until the next summer. After several more setbacks, it was successfully completed from Ireland to Newfoundland on 5 August 1858, to great rejoicing on both sides of the Atlantic.

Within a few weeks, however, the cable had failed, partly because of bad manufacture and handling and partly because of Whitehouse's use of very high voltages—including jolts from a five-foot induction coil. Even these massive jolts proved unable to operate Whitehouse's heavy relays, however, and few if any messages were sent successfully with his equipment. Thomson was brought in and had some success using batteries and his delicate mirror galvanometer, but the damage had already been done and by September the cable was effectively dead. Recriminations followed, with Whitehouse being given much of the blame. By the time the government-backed Red Sea cable to India failed in 1860 at a huge loss to the Treasury, long-distance cable telegraphy had acquired a reputation as a great *failed* technology. The British government and the Atlantic Telegraph Company undertook a joint investigation in 1859–60, and after taking expert testimony on both electrical and mechanical problems, the committee concluded that with more testing and better quality control, long cables *could* be laid and operated with confidence. In particular, the report endorsed Thomson's 'law of squares', backed up by extensive experiments by Latimer Clark, C.F. Varley, Fleeming Jenkin and other cable engineers. Retardation was still the focus of great interest; indeed, the committee declared that it remained 'the most important problem to solve in submarine telegraphy'.[54]

One important consequence of the effort to rationalize cable technology that followed the 1858 failure was the call issued at the 1861 meeting of the British Association for a standard set of electrical units. Scientists and telegraph engineers had been using various arbitrary resistance units for some years, and in 1860 Werner Siemens had introduced a mercury unit that achieved some currency but was not systematically related to other electrical quantities. Thomson's work on the Atlantic cable and in his Glasgow laboratory had convinced him of the practical and scientific value of standardized units, and he sought to promote the adoption of a system based on Weber's 'absolute' electrical units, which were directly related to units of force and energy. Thomson was unable to attend the 1861 British Association meeting (he was recovering from a badly broken leg), but he wrote to Wheatstone and other leading figures in the electrical world to seek support and relied on Fleeming Jenkin to manage the campaign at the meeting.[55] Amid this effort, Charles Bright and Latimer Clark arrived with a paper of their own proposing a system of standard units—the ohma, farad, galvat and volt—to enable cable engineers to compare and coordinate their measurements.[56] After some 'hot argument' on the question of whether Weber's units were the right size for practical use, Jenkin persuaded Bright and Clark to join the effort Thomson had already launched and to support the formation of what was initially called the 'British Association Committee on Standards of Electrical Resistance'. Over the next few years this committee—soon widened to cover all electrical units—established substantially the system of ohms, amps and volts that we still use.

The Standards Committee provides one of the clearest and most important examples of the fruitful interaction of cable technology and electrical science in the nineteenth century. It played a key role in opening up precision electrical measurement as a major area of electrical engineering and physics research in Britain. James Clerk Maxwell joined the committee in 1862 and, along with Jenkin and Balfour Stewart, did most of the experimental work on the determination of the ohm in his laboratory at King's College London. Cable engineers were mainly interested in having a unified set of electrical standards, but finding the value of the ohm in 'absolute' units was also important to them, since it would allow the charge and current involved in retardation to be directly compared.[57] This value was especially important to Maxwell, since his electromagnetic theory of light implied that the ratio between the electrostatic and electromagnetic units—proportional to the value of the ohm—was equal to the speed of light. Indeed, the agreement between the value of this ratio found by the Standards Committee and direct measurements of the speed of light long constituted the main empirical support for Maxwell's theory.

SUCCESS

Efforts to lay a cable across the Atlantic were put aside during the American Civil War, but in 1865 Field and his company tried again—this time following Thomson's advice much more closely than in 1858. The *Great Eastern* was leased and fitted out, and at first the laying went very smoothly. But about two-thirds of the way across the cable snapped, and it seemed that the great project had failed again. Field and his partners were able to raise money for yet another attempt in 1866, however, and this time they met with complete success; indeed, their engineers even managed to grapple up the 1865 cable, splice it and complete that as well. This success set off a great boom in submarine telegraphy, and by the mid-1870s an enormous network was in place, extending to India, Australia and the Far East—almost all of it controlled by British companies; indeed, mostly by one man, John Pender.[58] Besides forming a large and profitable business in itself, this cable network played a key role in facilitating the growth of the British Empire and the smooth operation of the world trading system that sustained Britain's wealth and power in the last quarter of the nineteenth century and the first quarter of the twentieth.

The growth of the cable industry had a profound and pervasive effect on experimental and theoretical work in electricity in Britain. As Maxwell noted in the preface to his *Treatise*, it created a 'demand for electrical knowledge' that had not existed before, and had the effect both of increasing the numbers engaged in electrical research and of 'stimulating the energies of advanced electricians' to study the new phenomena it presented.[59] In particular, cable telegraphy focused the attention of British electricians on propagation phenomena of just the kind Clark and Faraday had examined in 1853, and on the active role of the dielectric that Faraday had emphasized in his lecture at the Royal Institution.

Thomson brought these points out very clearly in his 1871 presidential address to the British Association at Edinburgh. After noting the contribution the work of the Electrical Standards Committee had made to Maxwell's electromagnetic theory of light, he said:

> This leads me to remark how much science, even in its most lofty speculations, gains in return for benefits conferred by its application to promote the social and material welfare of man. Those who perilled and lost their money in the original Atlantic Telegraph were impelled and supported by a sense of the grandeur of their enterprise, and of the world-wide benefits which must flow from its success; they

> were at the same time not unmoved by the beauty of the scientific problem directly presented to them; but they little thought that it was to be immediately, through their work, that the scientific world was to be instructed in a long-neglected and discredited fundamental electric discovery of Faraday's, or that, again, when the assistance of the British Association was invoked to supply their electricians with methods for absolute measurement (which they found necessary to secure the best economical return for their expenditure, and to obviate and detect those faults in their electric material which had led to disaster), they were laying the foundation for accurate electric measurement in every scientific laboratory in the world, and initiating a train of investigation which now sends up branches into the loftiest regions and subtlest ether of natural philosophy.[60]

This is as clear and sweeping a statement of my main points as I could ask for. Cable telegraphy, particularly the discovery of retardation, had focused attention on Faraday's 'long-neglected and discredited' ideas about the relationship between induction and conduction and the role of the dielectric, and so had helped to create a *market* for field theory—a market to which Faraday had himself laid claim in 1854. Moreover, that market exerted a strong influence on the later evolution of field theory, and I would suggest that it is to cable telegraphy, particularly the emphasis it placed on propagation phenomena, that we should look for clues to the direction British field theory took in the years after Faraday.

Acknowledgements
I would like to thank the Institution of Electrical Engineers, Trinity College Dublin, and the University Library, Cambridge, for permission to use manuscripts in their possession. Research for this paper was supported by NSF Grant 8911401.

Notes and References

The following abbreviations are used in the notes: *ER 3*, Michael Faraday, *Experimental Researches in Electricity*, vol. 3, (London, 1855); *E&M*, William Thomson, *Reprint of Papers on Electrostatics and Magnetism*, (London, 1872); *MPP 2* and *MPP 3*, William Thomson, *Mathematical and Physical Papers*, vol. 2, (Cambridge, 1884), and vol. 3, (Cambridge, 1890); *STP 1*, Werner Siemens, *Scientific and Technical Papers of Werner von Siemens*, vol. 1, (London, 1892); IEE, Institution of Electrical Engineers, London; ULC, University Library, Cambridge.

1. See J.L. Heilbron, 'The electrical field before Faraday', in G.N. Cantor and M.J.S. Hodge (editors), *Conceptions of Ether*, (Cambridge, 1981), 187–213.
2. Christa Jungnickel and Russell McCormmach, *Intellectual Mastery of Nature*, Volume 1: *The Torch of Mathematics, 1800–1870*, (Chicago, 1986), 120–1.
3. For a complementary perspective on the role 'technology' played in the origin of Faraday's field approach, see David Gooding, 'Experiment and concept formation in electromagnetic science and technology in England in the 1820s', *Hist. and Tech.*, 1985, 2: 151–76.

4. Michael Faraday, 'On the use of gutta percha in electrical insulation', *Phil. Mag.*, 1848, 32: 165-7, repr. in Faraday, *ER 3*, 494-6.

5. See the numerous items in the Airy Papers, ULC RGO 6/610, 611, 627, 633.

6. Latimer Clark to G.B. Airy, 13 October 1853, ULC RGO 6/468.173.

7. These experiments are described in a transcription of Clark's notes, IEE ScMSS 22/42-3, with a reference to the *Chemical Record*, 20 March 1852.

8. See the transcription of Clark's 'Memoranda of Experiments' in IEE ScMSS 22/22-41, and Faraday's notes in Thomas Martin (editor), *Faraday's Diary*, vol. 7, (London, 1936), 393-408.

9. Ibid. According to IEE ScMSS 22/26, the 15 October session was attended by Faraday, Airy, Edwin and Latimer Clark, 'and others'. These apparently included Latimer Clark's brother-in-law and assistant, W.H. Preece, later head of the Post Office telegraph department, who took some of the notes.

10. Martin, op. cit. (8), 398, 408.

11. 'A.B.G.' [G.B. Airy], 'Telegraphic longitude of Brussels', *Athenaeum*, 14 January 1854, 54-5; on Airy's authorship, see Airy to F.W. Evans, 6 January 1854 (pressbook copy), UCL RGO 6/634.235.

12. Michael Faraday, 'On electric induction—associated cases of current and static effects', *Proc. RI*, 1854, 1: 345-55, repr. in *Phil. Mag.*, 1854, 7: 197-208, and in Faraday, *ER 3*, 508-20.

13. Quoted from notes of the lecture apparently taken by W.H. Preece, IEE ScMSS 22/13-15. These notes differ in many ways from the published paper cited in the preceding note, which should not be regarded as simply the text of Faraday's lecture.

14. Faraday, *ER 3*, 508.

15. Latimer Clark, in Appendix 2 to the *Report of the Joint Committee on the Construction of Submarine Telegraph Cables*, British Parliamentary Papers 1860 [2744] LXII, 313.

16. On the Prussian lines, see Werner Siemens, *Inventor and Entrepreneur: Recollections of Werner von Siemens* (New York, 1966), 71-9, and Wolfgang Löser, 'Der Bau unterirdischer Telegraphenlinien in Preussen von 1844-1867', *NTM*, 1969, 6: 52-67.

17. Werner Siemens, 'Sur la télégraphie electrique', *Comptes rendus*, 1850, 30: 434-7, and *Annal. de Chimie*, 1850, 29: 385-430, trans. in Siemens, *STP 1*, 29-64. See also Michael Faraday, 'On subterraneous electro-telegraph wires', *Phil. Mag.*, 1854, 7: 396-8, repr. in Faraday, *ER 3*, 521-3, in which Faraday belatedly acknowledged Siemens's priority in discovering the 'jar-wire' effect. Few others in Britain gave Siemens's work even this much notice.

18. Werner Siemens, 'Ueber die electrostatische Induction und die Verzögerung des Stroms in Flaschendrähten', *Ann. Phys.*, 1857, 102: 66-122, trans. in Siemens, *STP 1*, 87-135, on 135.

19. Siemens later noted that Wilhelm Weber and other German theorists were unconvinced by his arguments and tried to explain his jar-wire phenomena in more orthodox ways, ascribing them either to electromagnetic induction or to electrical absorption; see Siemens, op. cit. (16), 163-5.

20. Löser, op. cit. (16), 59-61; Siemens, op. cit. (16), 89-90.

21. The British underground lines laid in the early 1850s also suffered from maintenance and insulation problems and were mostly replaced by overhead lines by the end of the decade—but by then the British were deeply involved in submarine telegraphy, in which there was, of course, no 'overhead' alternative.

22. Crosbie Smith and M. Norton Wise, *Energy and Empire: A Biographical Study of Lord Kelvin*, (Cambridge, 1989), 445-94, 661-83. While Smith and Wise

are certainly right to emphasize the large role telegraphy played in Thomson's thinking, I cannot agree with their claim that it lay behind his differences with Maxwell and the Maxwellians. There was no real conflict between adherence to Maxwellian ideas and a concern with telegraphy, as the work of Oliver Heaviside makes clear; indeed, his formulation of 'Maxwell's equations' owed much to his work on cable propagation. See my forthcoming paper, 'To rule the waves: cable telegraphy and the making of "Maxwell's equations" '.

23. See Edward Sabine to G.B. Airy, 9 December 1853 (extract), and Airy to Sabine, 19 December 1853 (pressbook copy), Airy Papers, ULC RGO 6/468.194 and 196. A transcription of the second of these is in the Kelvin Collection, ULC 7342-S8a; Smith and Wise, op. cit. (22), 456 n. 30, mistakenly give its date as 1854.

24. There is a preprint of Faraday's 1854 lecture in the Kelvin Collection, ULC 7342-PA412, but with no indication of when it reached Thomson.

25. [William Thomson], 'On the uniform motion of heat in homogeneous solid bodies, and its connection with the mathematical theory of electricity', *Camb. Math. Journal*, 1842, 3: 71–84, repr. in *Phil. Mag.*, 1854, 7: 502–15, and in Thomson, *E&M*, 1–14

26. William Thomson, 'On the mathematical theory of electricity in equilibrium', *Camb. and Dublin Math. Journal*, 1845, 1: 75–95, repr. (with additions) in *Phil. Mag.*, 1854, 8: 42–62, and in Thomson, *E&M*, op. cit. (23), 15–37.

27. William Thomson, 'On the electro-statical capacity of a Leyden phial and of a telegraph wire insulated in the axis of a cylindrical conducting sheath', *Phil. Mag.*, 1855, 9: 531–5, repr. in Thomson, *E&M*, 38–41, on 38.

28. See William Thomson to G.G. Stokes, 1 December 1854, Stokes Collection, ULC 7656-K75; cf. Smith and Wise, op. cit. (22), 453.

29. Abstracts of these papers are in the *BA Report*, 1854.

30. G.B. Airy to W.R. Hamilton, 19 September 1853, Hamilton Papers, Trinity College Dublin, MS 7767/1238.

31. William Thomson, 'Ether, electricity, and ponderable matter' [presidential address to the Institution of Electrical Engineers], *Journal of the IEE*, 1889, 18: 4–35, part repr. in Thomson, *MPP 3*, 484–515, quoted on 486.

32. Archibald Hamilton to 'Dixon', 6 January 1859, Hamilton Papers, Trinity College Dublin, MS 7767/1394; G.G. Stokes to William Thomson, 4 November 1854, Kelvin Collection, ULC 7342-S375.

33. Oliver Heaviside, *Electrical Papers*, (London, 1892), vol. 2, 79 and 86.

34. William Thomson to G.B. Airy, 19 January 1855, Airy Papers, ULC RGO 6/470.24–5; Thomson sent another letter on 20 January, 6/470.26–8, with curves showing the gradual rise of current.

35. G.B. Airy to C.S. Wollaston, 22 January 1855 (pressbook copy), Airy Papers, ULC RGO 6/470.29–31. Airy urged Thomson to come to London to see Latimer Clark's retardation experiments, but Thomson was unable to leave Glasgow; see Airy to Thomson, 22 and 26 January 1855 (pressbook copies), ULC RGO 6/470.32-5 and 38, and Thomson to Airy, 2 February 1855, ULC RGO 6/470.39–44.

36. C.S. Wollaston to G.B. Airy, 25 January 1855, Airy Papers, ULC RGO 6/470.36–7.

37. William Thomson to G.B. Airy, 2 February 1855, Airy Papers, ULC RGO 6/470.39–44. Thomson put the ratio of units between 104,000,000 and 419,000,000 cm/s; later determinations made it equal to the speed of light, about 300,000,000 cm/s.

38. William Thomson to G.G. Stokes, 12 February 1855, Stokes Collection, ULG 7656-K78.

39. William Thomson, 'On the theory of the electric telegraph', *Proc. RS*, 1855, 7: 382–99, repr. in *Phil. Mag.*, 1856, 11: 146–60, and in Thomson, *MPP 2*, 61–76.

40. William Thomson, 'On peristaltic induction of electric currents in submarine telegraph wires', *BA Report*, 1855, 21–2, repr. in Thomson, *MPP 2*, 77–8. The Black Sea cable rate is given in the *Joint Committee* report, op. cit. (15), xxiii.

41. E.O.W. Whitehouse, 'The law of squares—is it applicable or not to the transmission of signals in submarine circuits?', *Athenaeum*, 30 August 1856, 1092–3.

42. William Thomson, 'Telegraphs to America', *Athenaeum*, 4 October 1856, 1219, repr. in Thomson, *MPP 2*, 92–4.

43. William Thomson, 'Telegraph to America', *Athenaeum*, 1 November 1856, 1338–9, repr. in Thomson, *MPP 2*, 95–102.

44. Willoughby Smith, *The Rise and Extension of Submarine Telegraphy*, (London, 1891), p. 45.

45. Charles Bright, *Submarine Telegraphs: Their History, Construction, and Working*, (London, 1898; repr. New York, 1974), 54 n., states that his father, Sir Charles T. Bright, 'strongly urged' the use of a thicker cable but was overruled.

46. F.R. Window, 'On submarine electric telegraphs', *Proc. Inst. Civil Engineers*, 1857, 16: 188–202, with discussion, 203–25; quote on 200.

47. Ibid., 203.

48. Although Thomson's remarks were published as a letter to the Secretary of the Institution, they had in fact been read out by Bright in the course of his own remarks; see ibid., 210–11, and Charles Manby (Secretary of the Institution of Civil Engineers) to William Thomson, 21 January 1897, ULC 7342-I31. Thomson later said that he had 'assented too readily to [Whitehouse's] own over sanguine expectations of what his instruments could do', and that his remarks quoted in the discussion of Window's paper contained 'the most that I ever said in favour of the possibility of [Whitehouse's] success in getting a good rate'; see William Thomson to J.P. Nichol, 18 March 1859, ULC 7342-N30.

49. Michael Faraday, in discussion of Window, op. cit. (46), 221.

50. Cyrus Field and Alfred Varley, in discussion of F.C. Webb, 'Submarine telegraph cables', *Proc. Inst. Civil Engineers*, 1858, 17: 326 and 330–1.

51. Faraday, in discussion of Window, op. cit. (46), 222.

52. See, e.g., Bright, op. cit. (45), 23–56, 78–105, and Bern Dibner, *The Atlantic Cable*, (New York, 1959).

53. The chosen design was a little thicker than most earlier cables and Thomson regarded it as going some way toward answering his concerns, but he would have preferred a still thicker cable; see his December 1859 testimony, *Joint Committee* report, op. cit. (15), 111, and S.P. Thompson, *The Life of William Thomson, Baron Kelvin of Largs*, 2 vols, (London, 1910), vol. 1, 339.

54. Joint Committee report, op. cit. (15), xxiv.

55. Fleeming Jenkin to William Thomson, [undated], Kelvin Collection, ULC 7342-J36. Internal evidence indicates that these letters were written during the British Association meeting in September 1861 and that J37 was written a little before J36.

56. Latimer Clark and Charles Bright, 'Measurement of electrical quantities and resistance', *Electrician*, 1861, 3–4; an abstract, 'On the formation of standards of electrical quantity and resistance', was published in *BA Report*, 1861, 31: 37–8. See also Latimer Clark to William Thomson, 3 May 1883, Kelvin Collection,

ULC 7342-C91, quoted in Smith and Wise, op. cit. (22), 687.

57. See Latimer Clark and Robert Sabine, *Electrical Tables and Formulae*, (London, 1871), 1-2.

58. See Bright, op. cit. (45); Vary Coates and Bernard Finn, *A Retrospective Technology Assessment: Submarine Telegraphy—the Transatlantic Cable of 1866*, (San Francisco, 1979), 66-119, 156-79; Hugh Barty-King, *Girdle round the Earth: The Story of Cable and Wireless and Its Predecessors*, (London, 1979).

59. James Clerk Maxwell, *Treatise on Electricity and Magnetism*, 2 vols, (Oxford, 1873), vol. 1, vii-viii.

60. William Thomson, 'Presidential address', *BA Report*, 1871, lxxxiv-cv, repr. in Thomson, *Popular Lectures and Addresses*, vol. 2, (London, 1894), 132-205, quote on 161-2.

Telegraphy and the Technology of Display

The Electricians and Samuel Morse

IWAN R. MORUS

> If the presence of electricity can be made visible in any part of the circuit, I see no reason why intelligence may not be transmitted instantaneously by electricity.[1]

INTRODUCTION

The aim of this paper is to examine some aspects of the early genealogy of the electromagnetic telegraph by focusing on the implications of the only extensive piece of research that Samuel Morse ever published in a natural philosophical journal on the subject of his telegraph. This research, which was published in the form of a letter to *Silliman's Journal* in 1843, concerned a series of experiments that Morse had conducted using the equipment and apparatus assembled in preparation for the construction of the first telegraph line from Washington to Baltimore.[2] Work on this line commenced in September 1843 and was completed on 24 May in the following year. Some of the broader historiographical issues surrounding the relationship of science and technology will be addressed first in this paper.

Traditionally, historians both of science and of technology have conspired to present the relationship of science and technology in an unproblematic and hierarchical fashion. Historians of technology have taken 'science' to be an unproblematic source of natural facts which technologists then exploited for the production of material artefacts. Historians of science on the other hand have typically allowed no role for 'technology' in the production of scientific knowledge. The spread of new technologies has simply been seen as the application of accepted and verified knowledge. At best, historians (and philosophers) of science have regarded technological success as the final proof of a scientific theory's validity. Science has been regarded as being the product of creative and innovative thinking while technology has been relegated to the role of applying the products of that creativity.[3] The counter to this approach during the early 1970s was to argue that science and technology form autonomous and non-interactive discourses.[4]

More recently, however, hierarchical models of the science–technology relationship have been questioned by several commentators.[5] Drawing largely on the new sociology of scientific knowledge which was developed during the 1970s,[6] many historians and sociologists were led to question the validity of any hard and fast distinction between science and technology. The result has been the emergence of several 'interactive' models of science and technology which emphasize the fluid and ever-changing relationship of the two categories and also bring to the fore the importance of politics, economics, industry and other 'social' factors in the development and diffusion of new technologies.

Thomas Hughes, in his account of the introduction of electrical power systems at the close of the nineteenth century, emphasizes what he calls the 'seamless web' of scientists, inventors, engineers, industrialists and politicians who had a role to play in the introduction of the new system.[7] Hughes's main point is that the interaction between these different groups was not hierarchical but that they were all equally important components of the system and had equally important roles to play in its eventual success. He emphasizes that as a result of the fundamental interconnectedness of these groups, any comprehensive history of a new technological system would require an account of *all* these different groups and individuals and their interactions.

A similar approach is advocated by Michel Callon and John Law, who also emphasize the heterogeneity of networks.[8] Both these commentators analyse technological systems as networks of a variety of heterogeneous actors, both human and non-human. Networks consist of entities such as knowledge, artefacts, persons and groups, which are organized and orchestrated in order to solve particular problems and achieve particular goals. Once again, one important implication of such an analysis is that networks are not organized hierarchically. All the different components of the network are equally crucial to the successful outcome.

Trevor Pinch and Wiebe Bijker approach the problem from a slightly different perspective.[9] They recognize a distinction between science and technology but reject the notion that the distinction is hierarchical. They maintain that the distinction is the result of historically and socially situated negotiation and that it is, as a result, always open to challenge and reconstruction. Such an analysis is not at all incompatible with the claims of the systems or network analysis since it accepts that the attribution of labels such as 'science' or 'technology' to particular items of research is invariably retrospective and has no role to play in the actual construction of any particular historical episode.

The account I offer here of the early history of the telegraph is not intended as an exemplar of any one of these approaches, although it is, I hope, informed by all of them. My aim is first of all to reconstruct the original historical context in which the techniques and the material artefacts utilized by the early telegraphers were developed. I will claim that, in both the UK and the United States, the evidential context within which electrical techniques and artefacts were developed was situated within a culture which privileged display and variety rather than measurement.[10] It was this culture that originally made available the technology of the early telegraph, and the first telegraph inventors were themselves part of this culture.

Having delineated the contours of this electricians' culture I will then focus in detail upon Samuel Morse's telegraph and the research paper mentioned previously. The aim here will be to highlight some of the problems facing men such as Morse in taking the artefacts and techniques of electricity out of their original settings in the laboratory and the lecture theatre and into the real world. I will also point out some of the strategies adopted by Morse in his paper to achieve that goal.

I do not wish to claim that such an account constitutes a comprehensive history of the development of the telegraph. I will make no effort, for example, to discuss the vast amount of political lobbying in which Morse was obliged to engage between 1835 and 1842 in order to secure financial backing for his project. Such a discussion would be an essential part of any fully rounded history. My account is intended as a corrective to the traditional view that the telegraph, for example, was simply based on the application of principles developed by 'scientists' such as Michael Faraday and that its success was a straightforward vindication of the truth of those principles. I will argue that the telegraph emerged from its own distinctive culture and that its success was contingent upon the ability of its inventors to take it out of the laboratory and into the world.

CIRCUS AND CIRCUITS

By the mid-1830s the study and display of electrical phenomena were firmly established in the experimental repertoire of English natural philosophers, particularly in the wake of Oersted's experiment of 1820, which established the long-sought-for link between electricity and magnetism. The major metropolitan centre for electrical research was the Royal Institution, where Michael Faraday continued in the tradition established by Sir Humphry Davy of combining philosophical discovery and public edification. Faraday and others, such as John Frederic Daniell, constituted a self-proclaimed élite of the London electromagnetic network.[11] Their position as an élite was not purely constituted by either social class or intellectual superiority. Rather it was a consequence of their entrenched positions within secure institutions which both provided them with resources and endowed their pronouncements with intellectual authority.[12] The social network surrounding the Royal Institution in particular guaranteed that pronouncements on electrical matters by Michael Faraday would gain rapid and authoritative circulation within the higher echelons of London's philosophical community.[13]

This élite was not, however, the only component of the electrical network of this period. If Faraday and his fellow members of institutions such as the Royal Institution may be characterized as practitioners of a professorial and gentlemanly science, then there also existed a set of actors who might be characterized as practitioners of a populist electricians' science. The lecturer William Sturgeon is a good example of this other culture.

By the mid-1830s Sturgeon was well-established as a public lecturer on natural philosophical subjects and a philosophical instrument-maker. He had been born the son of a shoemaker in 1783 and after a short apprenticeship in that trade volunteered for the Royal Artillery in 1804. Upon leaving the service in 1820 he settled at Woolwich near London and supplemented his income as a shoemaker by lecturing on natural philosophy. In 1824 he was appointed lecturer in science and philosophy at the East India Company Royal Military College at Addiscombe. From 1832 Sturgeon was also on the lecturing staff of the Adelaide Gallery of Practical Science near the Strand in London.

One of the major difficulties facing actors such as Sturgeon was that the successful display of electrical and electromagnetic phenomena required the use of apparatus that was both expensive and difficult to manipulate before an audience. This problem did not present itself as urgently to those such as Faraday who had access to the facilities and resources of the Royal Institution. One of Sturgeon's major concerns during the 1820s was therefore the search for a means of producing such phenomena as the Oersted effect conveniently and economically. This was the interest that underlay his invention of the electromagnet, for which he was awarded a silver medal by the Society of Arts in 1825.[14] This concern with display and practical economy was to structure the work of many electricians over the next two decades.

The Adelaide Gallery where Sturgeon occasionally lectured was a particularly important centre for the electricians' community during the 1830s. It was founded in 1832, by a set of interests ranging from the wealthy philanthropist Ralph Watson to the engineer Thomas Telford, as the *National Gallery of Practical Science, Blending Instruction with Amusement.* Its object was:

> to promote . . . the adoption of whatever may be found to be comparatively superior, or relatively perfect in the arts, sciences, or manufactures . . . [and to display] subject to immediate return on demand, and meanwhile protected by every precautionary arrangement, specimens and models of inventions and other works &c. of interest, for public exhibition, free from charge . . . thereby gratuitously offering every possible facility for the practical demonstration of discoveries in Natural Philosophy, and for the exhibition of any new application of known principles to mechanical contrivances of general utility.[15]

Men such as Sturgeon took full advantage of this facility to display their apparatus to the public, who were charged a shilling for the opportunity of examining electrical machines and other curious contrivances such as the steam gun. Those who visited the Gallery were prepared to witness wonders:

> Clever Professors were there, teaching elaborate science in lectures of twenty minutes each. Fearful engines revolved and hissed and quivered. Mice led gasping sub-aqueous lives in diving bells. Clockwork steamers ticked round and round a basin perpetually to prove

> the efficacy of invisible paddle wheels. There were artful snares laid for giving galvanic shocks to the unwary.[16]

This was the context within which popular electricians chose to work during the 1830s and 1840s in London.

A survey of papers submitted by electricians to scientific journals, such as the *Philosophical Magazine*, provides some interesting insights into their concerns. Many of the contributions were concerned with the construction of electrical machines and instruments, and were frequently little more than advertisements for a particular electrician's latest model. The interesting point about such contributions, however, is that they typically took for granted that the function of the instruments they described would be primarily one of straightforward display. They were perceived by their makers as being designed for the lecture theatre rather than the laboratory. Or, more accurately, their makers took for granted that a technology of display was as appropriate in the laboratory as it was in the lecture theatre.

An example should make the point explicit. In 1836 the instrument-maker E.M. Clarke published an account in the *Philosophical Magazine* of what he claimed as an important new version of the electromagnetic machine.[17] The account consisted of a description of the machine and an exhaustive listing of its advantages compared with other versions. Clarke's list drew attention to the variety of *experiments* the machine could perform:

> By my intensity armature . . . I am enabled to go through the various experiments that are usually performed by a number of separate galvanic plates. The effect this armature produces on the nervous and muscular system is such, that no person out of the hundreds who have tried it could possibly endure the intense agony it is capable of producing. . . . It shows the decomposition of water . . . and also of the neutral salts; it deflects the gold leaves of the electroscope, charges the Leyden Jar, and by an arrangement of wires . . . the electricity is made distinctly visible, passing through the magnetic battery to the armature, and by the same arrangement not only shocks, sparks, but also brilliant scintillations of steel can be obtained.[18]

The crucial feature of this account is the function accorded to the word *experiment*. As used by Clarke the term acquires meaning by reference to a technology of display within which an experiment is not so much regarded as a means of discovery but as an item of theatre.

This much appears unproblematic, but the controversy which immediately surrounded this text showed clearly that the same status could be accorded by the network to experiment as display as was accorded to experiment as discovery. No clear line was drawn between the discourse of the theatre and the discourse of the laboratory. The following issue of the *Philosophical Magazine* contained a response to Clarke from a rival instrument-maker, the American Joseph Saxton, in which Clarke's claims of novelty for his machine were strongly disputed. Crucially, Saxton used the term 'experiment' in exactly the same way as Clarke. In other words, he too referred experiment to a technology of display.[19]

The exchange does make explicit the point that the technology of display associated with such instruments as the electromagnetic machine could not be dissociated from the discourse of experimental discovery. Both Saxton and Clarke were in agreement on the importance to be attached to claims of priority in the production of such displays and did not question the validity of according them the status of experiment. Indeed it was precisely the experimental status accorded to such productions that rendered it essential that priority should be correctly attributed.

The differing interests of the electricians and men such as Faraday or Daniell are particularly apparent in their different conceptions of an item of apparatus—the voltaic battery—and the repertoire of effects and instruments associated with it.[20] Daniell, who invented the first constant voltaic cell in 1836, simply perceived the battery as an unproblematic source of steady electrical output: a black-boxed electrical instrument rather than a site for further philosophical and experimental enquiry in its own right.

The electricians, on the other hand, constructed a whole repertoire of techniques surrounding the battery. Their aim was to make electricity visible on a grand and spectacular scale and for this purpose they exploited the whole range of the battery's powers, from its capacity to cause shocks to its capacity to produce light effects. The following extract from a paper by the electrician Alfred Smee is typical:

> Four cells, containing 48 square inches in each cell, decomposed 7 cubic inches of mixed gases per five minutes, whilst four cells of Professor Daniell's, in which 65 square inches of copper were exposed in each cell, gave off only five cubic inches in the same time. . . . This battery also possesses great heating powers, raising the temperature of a platinum or steel wire, 1 foot long and of a thickness similar to that used for ordinary birdcages, to a heat that could not be borne by the finger. Its magnetic power is not less astonishing, three cells supporting the keeper of a magnet through forty-five, two cells through thirty-two, and one cell through twenty thicknesses of paper.[21]

Such accounts invariably emphasized the variety and the grand scale of electrical effects.

For members of the elite, such as Faraday, displays of the battery's power were to be treated as unproblematic measurements of electrical action. Such a concern is evident in Faraday's work on the volta-electrometer in which a display of the battery's power of chemical decomposition was transformed into a measurement of its output. Faraday's claims for this instrument were impressive:

> The instrument affords us the only actual measurer of voltaic electricity which we at present possess. For without being affected by variations

> in time or intensity, or alterations in the current itself, of any kind, or from any cause, or even intermissions of action, it takes note with accuracy of the quantity of electricity which has passed through it, and reveals that quantity by inspection. I have therefore named it a VOLTA-ELECTROMETER.[22]

Faraday's interest in black-boxing the volta-electrometer as an unproblematic item of laboratory apparatus stood in stark contrast to the concerns of William Sturgeon, who strenuously attacked Faraday's interpretation of his experiments, seeing the volta-electrometer (or the 'electro-gasometer' as he called it) as one of a variety of displays rather than a unique measurement.[23]

It should be emphasized that this was not a straightforward dispute between quantitative and qualitative scientists. Both sides in the dispute were interested in quantification but differed as to what was being quantified. Faraday was interested in using the decomposing power of the battery as a means of measuring its electrical output. Sturgeon, on the other hand, was interested in quantifying and measuring the different powers of the battery in order to determine which type of battery was best suited for which kind of display.[24] The dispute makes clear the different interests of different sections of the network. The electricians were interested in instruments as constituting an economical technology of display, while Faraday was more interested in instruments as means of providing surrogates for electrical action. The difference was between a culture that regarded experiment as theatre and one that regarded experiment as a tool for interrogating an external nature.[25]

The theatricality of the electricians' culture is underlined by events such as the 'electrical soirées' held by the wealthy electrician John Peter Gassiot to entertain foreign *savants* when they visited London. At these events electrical experiments would be carried out in front of the guests on a grand scale. Gassiot, as a wealthy wine merchant, had access to resources which few of his electrical contemporaries could match and during the late 1830s and early 1840s his laboratory was probably the best equipped for display in London.

Following his visit to London during the early 1840s, the French-Swiss electrician Auguste de la Rive presented a report to the Geneva Philosophical Society which was translated in C.V. Walker's *Electrical Magazine*, describing the state of the study of electricity in England. Gassiot had held an electrical soirée in his honour while de la Rive was in London, and de la Rive was sufficiently impressed to devote the second half of his account to a description of that event:[26]

> I had the pleasure of assisting at these experiments, or rather at this interesting *electrical soirée* as it was called by the amiable *savant*, who had been so kind as to open his house, with the most cordial hospitality, to the numerous assistants who crowded forward to see and admire. A Grove's battery of 100 pairs (each presenting a surface of about 16 sq. in.) produced magnificent effects; the luminous arc, between the charcoal points, lighted up a very large apartment,

> and made all the wax lights become pale, for their light almost entirely disappeared before that of the electric current; wires nearly a line in diameter were made red hot to the length of more than a foot.[27]

Once again it is worth noting the meaning of *experiment* in such a context.

Gassiot held a similar soirée when the Italian electro-physiologist Matteucci visited the country in 1845. Walker's *Magazine* again reported the event:

> On Friday, Oct. 4th, in compliment to Prof. Matteucci, Mr. Gassiot held an Electrical Soirée at his house at Clapham, to which a large party of the Professors and lovers of science were invited to meet the illustrious philosopher of Pisa. The suite of rooms were thrown open at 8 o'clock; the tables were covered with electrical apparatus, and the results of electricity; among which latter we observed a very fine salver of solid silver, and a large copper electrotype. Mr. Gassiot several times during the evening satisfactorily repeated his experiment of tension before contact. His series of 100 Grove's was excited; wires were carried out upon the lawn, where the charcoal light was exceedingly brilliant,—so much so that its reflection was seen at the distance of several miles, and the appearance was believed to be due to some atmospheric meteor.[28]

Obviously, such events were quite explicitly conceived of as simple entertainment. However, they merged very easily with far more 'scientific' activities.

Gassiot's own first contribution to the *Philosophical Transactions* of the Royal Society is a case in point. In this paper Gassiot described several experiments he had conducted in an attempt to produce an electric spark from a large battery before the circuit was completed. Such a phenomenon was regarded as the final proof required of the identity of galvanic and frictional electricity. Faraday had claimed in 1834 that he had succeeded in producing this phenomenon but Professor Jacobi in St Petersburg had disputed the validity of Faraday's results in 1838.[29]

Gassiot's first task in the paper was to convince his audience of the power of his apparatus. This was of particular importance since he had not, in fact, succeeded in producing the spark before contact. His paper was therefore a potential refutation of Michael Faraday's claims. Gassiot described the effects of his apparatus as follows:

> Those who had the pleasure of witnessing the experiments of Professor DANIELL, at King's College, when a series of only seventy pairs of his constant battery was used, will no doubt recollect the brilliant effects produced with this powerful apparatus, and may form some idea of those obtained using nearly one third increase of power. Titanium, which had been previously given to me by Dr. FARADAY, was fused into a solid mass, and is now in that gentleman's possession; platinum was volatilized; and the flame from charcoal as well as from metallic electrodes was so intense as to render it indispensable that the eyes of those present should be protected by thick screen of

> black crepe. On another evening sixteen feet four inches of platinum wire No. 20 gauge, was ignited to a red heat; and even this length might have apparently been extended had I a greater quantity of that wire; but with such a powerful apparatus no effect [i.e. no spark before contact] could be obtained even through one thickness of a silk handkerchief.[30]

The implication of this passage is that theatre was an integral part of experimental life for men such as Gassiot. Even an account of a private experiment in a private laboratory was to be couched in the language of display.

It is also worth noting the use of witnesses in this passage. It is evident that Gassiot's experiments were carried out in the presence of witnesses, and it is no coincidence that he chose to cite Faraday and Daniell to legitimize his experiments as they were among the most eminent of London's natural philosophers. Experimenting in the presence of eminent witnesses gave authority to the results: by linking his name to such figures a less established electrician, such as Gassiot at this early stage in his career, could add credibility to his own claims by appealing to the sanction of *their* credibility. The presence and support of good witnesses guaranteed the credibility of Gassiot's own experimental techniques and his practical competency.[31]

THE AMERICAN SETTING

There is as yet no comprehensive survey covering the community of electrical practitioners in the United States in this period. There is, however, some evidence to suggest that, certainly during the 1830s and early 1840s, institutional factors there led to the existence of a network of electricians with much the same kind of interests and practices as have been described for the English situation in the previous section. Many American scientists of the period were, for example, employed as lecturers and professors at the various colleges and academies of the East Coast, such as Yale, Harvard and New Jersey (later Princeton) Colleges.[32] Their main duty at these institutions was to teach and as a result many of their scientific interests were centred around the production of techniques and apparatus for lecturing and demonstrating.

The East Coast by the early nineteenth century also maintained a significant number of instrument-makers, particularly in the Boston and Philadelphia areas. Joseph Saxton of Philadelphia, for example, had originally been apprenticed as a clockmaker before embarking upon a career as a scientific instrument-maker. For most of the 1830s, Saxton was resident in London, where he designed and built many of the exhibits for the Adelaide Gallery and also constructed the apparatus used by Charles Wheatstone to measure the velocity of the electric fluid.[33] Daniel Davis of Boston was the first American instrument-maker to specialize in the construction of electrical apparatus. He had close links with Charles Grafton Page and constructed many of his instruments.

Such men had a concern for technologies of display and means of making the phenomena visible in much the same way as their English counterparts, despite the differences in institutional setting. It is also worth noting that men such as Benjamin Silliman, James Freeman Dana and Joseph Henry had frequent contacts with English electricians. Both Silliman and Dana visited Britain very early in their careers, having been sent over from the United States to acquire scientific instruments for teaching purposes by their respective institutions.[34] The Joseph Henry papers also contain frequent items of correspondence, either from or concerning English electricians. Henry had visited England in 1837.[35]

In fact Henry's early work in electricity shows very clearly the extent to which the technology of display was central to his practice as a scientist during the late 1820s and early 1830s. Henry first started working on electricity and electromagnetism after his appointment as Professor of Mathematics and Natural Philosophy at the Albany Academy about 1826. Significantly, his first contribution was explicitly presented as an extension of William Sturgeon's recent researches on the electromagnet and the construction of economic and efficient apparatus for the production of display.

Henry was obviously perfectly well aware of the significance of Sturgeon's work and the interests underlying it. Discussing the science of electromagnetism in the introduction of his first paper on the subject, read before the Albany Institute on 10 October 1827, he drew attention to the demonstrative potential of the new science: 'Our popular lecturers have not availed themselves of the many interesting and novel experiments with which [electromagnetism] can so liberally supply them; and, with a few exceptions, it has not as yet been admitted as a part of the course of physical studies pursued in our higher institutions of learning.'[36] He then went on to suggest that the principal reason for this lack of interest was the high cost of constructing and maintaining electrical apparatus that was sufficiently powerful to display electromagnetic phenomena to good effect. Sturgeon's work, he claimed, overcame this difficulty.

Henry's own 'modifications of the electro-magnetic apparatus' were presented as extensions of Sturgeon's work, suggesting means of constructing apparatus which could display on a large scale various phenomena that could not be produced using Sturgeon's apparatus. In particular his apparatus was designed to demonstrate the action of the earth's magnetism on galvanic currents and the operation of adjacent current carrying wires on each other. In subsequent papers, also written while Henry was Professor at the Albany Academy, he extended Sturgeon's work on the electromagnet itself, showing how by winding the coils of wire very tightly about the iron core, a massive increase in magnetic power could be achieved.[37]

It should be noted that this work by Henry was to be crucial for the success of Morse's telegraph a decade later. It was only when Morse discovered Henry's researches through the agency of his friend Leonard Gale, then Professor of Chemistry at the University College of New York, that he succeeded in producing electromagnets that were of sufficient strength to operate the telegraph through long distances of wire. Henry himself had not been unaware of the potential of his work as an avenue towards the construction of some kind of communication device but he made no attempt to pursue the idea beyond the laboratory.[38] One of the assertions of this paper will be that one way of describing Morse's achievement is in terms of his success in taking Henry's experiments and technology of display and making them work outside the laboratory.

Charles Grafton Page's work during the 1830s also provides evidence of the American electricians' concern with technologies of display.[39] Page had graduated from Harvard in 1832 and received his degree of MD in 1836. Following the completion of his training at Massachusetts General Hospital he set up his medical practice in his home town of Salem. His training at Harvard had included courses in chemistry and physics, both of which included extensive laboratory demonstrations. His teacher there had been John White Webster, the Erving Professor of Chemistry and Mineralogy, who was later hanged for murder in 1850.[40]

One of Page's first papers: 'Method of increasing shocks, and experiments, with Prof. Henry's apparatus for obtaining sparks and shocks from the Calorimotor', shows his affinity with the technology of display quite clearly.[41] In this paper he extended Henry's recent work on induction, showing that it was possible to produce shocks and sparks from sections of the coil exterior to the battery circuit. As Post points out, this paper and others by Page from this period were primarily concerned with the construction of apparatus for the spectacular demonstration of new electromagnetic phenomena.[42] The concern is evident, for example, in his papers on the production of sound by magnetism, a phenomenon which Page named 'galvanic music'.[43] William Sturgeon was certainly sufficiently impressed by his work to republish a large proportion of Page's papers in his *Annals of Electricity*.[44]

TELEGRAPHY AND DISPLAY: THE COMMON CONTEXT

It is the main contention of this paper that, certainly in England and the United States, both telegraphy and telegraphists emerged from this common culture and its technology of display. The electricians' culture, with its emphasis on variety and its interest in making electricity visible in all its manifestations, constructed a wide array of practical, economic and efficient techniques and devices for achieving such ends. They were also interested in the utility of natural knowledge. Early telegraphic devices were constructed by members of the electricians' community out of this array of apparatus for displaying electricity. The success or failure of the various telegraphs that were patented during the late 1830s and early 1840s would be contingent upon the ability of their inventors to take them outside the laboratory. The electricians' culture which embraced the laboratory and the lecture theatre would have to be extended to include the rest of the world.

The basic principle of the telegraph was quite straightforward. The apparatus consisted of a galvanic battery, a circuit breaker and some means of registering the presence of electricity in the circuit—in other words any one of the wide variety of instruments devised by the electricians to make the power of the battery visible. By the repeated breaking and completing of the circuit a display would be continually produced and reproduced by the instrument, and by variance of the pattern of circuit breaks a message could be constructed.[45]

One of the first successful telegraphs devised in England was that invented by Charles Wheatstone and William Fothergill Cooke, for which a patent was given in 1837. The basic item of the technology of display utilized in this telegraph system was the galvanometer. In the original Wheatstone and Cooke patent, an array of five galvanometer needles was connected by a five-wire circuit to a transmitting device which consisted of five keys that acted as switches. By the closing and opening of different combinations of these keys, the array of galvanometer needles could be made to arrange themselves in different patterns of left and right positions so as to indicate different letters of the alphabet. A message could thus be transmitted down the line.[46]

Later in 1837 the instrument was tested over a 1.4-mile line running along the main London to Birmingham railway between Euston Square and Camden Town. The experiment was successful in the sense that messages could be transmitted, but as a result of mechanical difficulties in laying the lines, the difficulty of ensuring that the wires were sufficiently insulated and the expense of the project, the method was not pursued further. However, simplified and improved versions of the basic galvanometer principle were soon devised by Wheatstone and Cooke. These proved more successful and were used in practice by the mid-1840s.[47]

Other British telegraph inventors utilized other items of the technology of display. Both Edward Davy and Alexander Bain used the electrochemical decomposing powers of the galvanic battery to construct a telegraph system. In Davy's apparatus, which was patented in 1838, galvanometer needles were activated (as in the Wheatstone and Cooke system) but rather than being used as receivers they were used to close a local circuit which passed a current through a strip of chemically treated fabric. Chemical decomposition of the fabric took place producing discoloration, and by repeatedly opening and closing the circuit in different patterns a coded message could be recorded on the chemically treated tape. However, Davy failed to persuade any investors to provide him with the financial backing required to test his invention outside the laboratory.

Alexander Bain was more successful. In his telegraph, a tape of chemically treated paper was passed through a break in the circuit. As the circuit was opened and closed according to a set pattern at the transmitting end, a current would pass through the treated paper producing discoloration in a sequence of dots and lines which could be used as a code.[48] Unlike Davy, Bain succeeded in finding financial backing and his system was used on a practical basis in both Britain and the USA by the end of the decade.

In the long term the most successful telegraph inventor was Samuel Morse. For most of the nineteenth century the origins of the Morse telegraph system were surrounded by controversy. During the few decades following his first successful experiments Morse quarrelled with most of his partners and associates and as a result found himself embroiled in a series of priority disputes concerning who could lay claim to the actual invention of the system.

According to Morse's own account, he first became aware of the possibility of using electricity as a medium for communication as the result of a conversation aboard the packet-ship *Sully* during his return voyage at the end of a European tour in 1832.[49] He already had some knowledge of electricity as a result of his friendship with Benjamin Silliman, with whom he had spent some time carrying out experiments on electricity and magnetism during 1820 and 1821.[50] He had also attended a course of lectures on the subject given by his friend James Freeman Dana at the New York Athenaeum in 1827. Dana was then Professor of Chemistry at the College of Physicians and Surgeons at New York.

Morse was unable to pursue his ideas concerning the telegraph until after his appointment as Professor of the Literature of Art and Design at the University College of New York in 1835. Shortly after his appointment he succeeded in building a crude working model operating by means of an electromagnet. The transmitter consisted of a printer's composing stick, which was passed through a device for opening and closing the circuit according to the indentations of the stick. The receiver consisted of an electromagnet, which as it was activated and deactivated by the opening and closing of the circuit moved a stylus that inked zigzag marks on a paper tape. In later versions of this apparatus the stylus was lifted and dropped by the electromagnet rather than being moved back and forth, so that a series of dots and lines were inked on to the tape.

Morse was then faced with the usual problem of the telegraph inventor: making this item of the electricians' technology of display operate outside the laboratory or the lecture theatre. His first model would only operate through less than forty feet of wire. He turned for assistance to Leonard Gale, the Professor of Chemistry at the University College of New York. Gale was familiar with Joseph Henry's work on the construction of powerful and economic electromagnets. He and Morse spent the summer and autumn of 1837 rebuilding the telegraphic apparatus using Henry's work and by the end of November 1837 could operate their telegraph through ten miles of wire. This was all that they felt was required since Morse had devised a relay that could repeat any signal from one ten-mile circuit to the next.

With the practical and financial backing of Alfred Vail, with whom Morse and Gale entered into partnership (with Morse as the senior partner), a working model of the telegraph was constructed for which Morse applied for and received a caveat on 3 October 1837. During the first months of 1838 the new instrument was publicly demonstrated on several occasions, culminating in a demonstration on 20 February before the Committee of Commerce of the House of Representatives. F.O.J. Smith, the chairman of the committee, was sufficiently impressed to join the partnership. He and Morse spent the rest of 1838 and much of

1839 in Europe, unsuccessfully seeking foreign financial support for the new invention. A US patent for the telegraph was finally granted on 20 June 1840 but it required three years of intensive lobbying before Morse and his partners eventually secured a grant of $30,000 from Congress, which they used to construct an experimental line from Washington to Baltimore along the Baltimore and Ohio Railroad.

Before work commenced on the laying of the line Morse used the accumulated equipment to carry out a number of experiments. The resulting paper was published in *Silliman's Journal* for September 1843.[51] In the closing section of this paper some of the implications of this publication for the early history of telegraphy will be considered.

EXPERIMENTS WITH GROVE'S BATTERY

The experiments carried out by Morse on 8 August 1843 were by no means original or unusual. They differed from the standard repertoire of the electricians' culture only in terms of the scale of the apparatus and the resources at Morse's disposal. Morse described the experiments as being carried out in verification of the law of Lenz concerning the action of galvanic electricity through wires of great length.[52] According to this law, the quantity of electricity which could pass through a series of wires whose lengths were in arithmetical ratio would decrease in a geometrical ratio.

The source of galvanic electricity used by Morse was a battery of 100 pairs of the nitric acid cell recently invented by the British electrician William Robert Grove. The Grove cell was by far the most powerful battery available at this time, its main drawback being that one of its plates was of platinum, which made the combination comparatively expensive to construct.[53] Morse, because of his grant from Congress, did not need to worry unduly about the expense of his electrical power source and could therefore choose the battery that would display his apparatus's viability to the best advantage. The wire for the experiments was separately coiled on eighty reels in units of two miles per reel so that Morse could easily vary the length of his circuit by adding more reels of wire.

The first experiment conducted by Morse was simply to introduce his electromagnet into the entire circuit of 160 miles. He noted that with all the Grove cells providing the power source the magnetism induced was 'sufficient to move with great strength my telegraphic lever'. When the number of Grove cells included in the battery was decreased the electromagnet continued to attract the lever down to a combination of forty-eight cells.

Morse then proceeded to test the battery's decomposing power at various distances and gave the results in inches of gas produced per minute from a number of combinations ranging from the battery alone with no reels of wire interposed up to fifty miles of wire interposed in the circuit. He noted that during the previous summer, using smaller apparatus, he had attempted a similar set of experiments using electromagnetism as a measure of the battery's power. This would have been an obvious test to employ since the telegraph itself operated by electromagnetism. He had, however, been unable to measure the lifting power of the magnet at long distances with sufficient accuracy, which presumably accounts for his choice of a different measuring technique on this occasion. He could now claim retrospectively that the results produced during the previous summer were in sufficient agreement with the present results to confirm Lenz's law.[54]

The amount of preparatory work carried out by Morse in anticipation of this experimental session provides an interesting indication of the extent to which he operated within the electricians' culture of witnessing and display. As suggested previously, electricians regarded the presence of authoritative and competent witnesses as an essential prerequisite for the conduct of a good experiment. Morse went to a great deal of trouble to try to ensure the presence of such witnesses. He wrote to Joseph Henry on several occasions, urging him to attend. He even supplied him with details of the stagecoach journey from New York to the site of the experiment:

> We have delayed a little in our arrangements, but on Tuesday the 8th of August (one week from tomorrow) I shall leave my office in Nassau Street at 9 o'clock A.M. in the Bloomingdale Stage for Mr Chase's ropewalk at the junction of 8th avenue and the Bloomingdale road, about 6 miles from the city Hall. I hope that Professor Torrey & yourself will be able to go out with me at that time. If by any accident you should not be with me in season, you can reach the same spot by the *Knickerbocker* stages, which are passing Broadway every few minutes; They go as far as 21st Street, whence any half hour they proceed to Manhattansville on the 8th Avenue, passing the above mentioned ropewalk. I am expecting quite a reunion of the scientific men in our vicinity, Prof[rs] Silliman, Renwick, Ellet, Draper &c. &c.[55]

In the event Henry, Torrey and Silliman were all unable to be present. The experiment was, however, witnessed by James Renwick, Professor of Natural Philosophy and Experimental Chemistry at Columbia University, John William Draper, Leonard Gale's successor as the Professor of Chemistry at the University College of New York, Professor Ellet and G.C. Schaeffer. Morse's two assistants, Fisher and Leonard Gale, were also present.

Two observations may be made at this point. The immediate objects of Morse's experiments are self-evident. He was using the oppportunity both to publicize his work on the telegraph within the natural philosophical community and to demonstrate its philosophical possibility. It was essential that both these aims should be achieved. Morse's future success in gaining funds from the Congress would be contingent upon the continued support of the philosophical community, which would have to function as sponsor in future representations to the government.

Equally important was the need to demonstrate the telegraph's viability. Potential critics familiar with the literature on electricity would have no difficulty in finding authoritative denials of the theoretical possibility of an electric telegraph. The English electrician Peter Barlow, for example, had carried out a series of experiments on long circuits as early as 1825 and had come to the conclusion that the magnetic effect of a current carrying wire diminished as the square root of the distance along the wire from the battery.[56] This claim from what was a very authoritative source had not been theoretically refuted. Various commentators had argued that the difficulty could be overcome by simply increasing the size of the apparatus, but such an option was not necessarily economically feasible.

If the results presented by Morse were accepted—and Morse had made every effort to secure the necessary authoritative witnesses who would guarantee the competency of his experimental practices—then Barlow's theoretical objections could be overcome. Appended to Morse's letter was a calculation supplied by Morse's colleague at the University College of New York, John Draper.

He represented Lenz's law as the logarithmic equation $x = a^y$, where x was the quantity of electricity, y the length of the wire and a a constant representing the resistance of a particular wire. The relevant section of the curve lies in the fourth quadrant, with x positive and y negative. When y approached zero, the quantity of electricity would approach unity and as it increased the quantity of electricity would diminish towards zero. The crucial point was that as y was increased by successively larger amounts the diminution in electricity would become smaller until eventually, a doubling for example, of the circuit's length would produce no appreciable diminution in electrical effect. The implication of this for Morse's telegraph was clear. If Morse could show, as he had, that his telegraph could operate through 160 miles of wire, he could plausibly argue that it would operate as well through 300 or even 600 miles.

Briefly then, Morse published his results in *Silliman's Journal* as a public relations exercise. To a certain extent such publication was a risky strategy since it could have made some of the specifications of his telegraph equipment available to potential competitors in an unprotected context. On the other hand, it gave him the opportunity to alert the natural philosophical community to his telegraphic work and its potential scientific interest and it allowed him to display the theoretical possibility of taking his technology of display out of the laboratory and into the world. In order to achieve these aims it was essential that Morse's work be recognized as belonging to the electrician's genre: hence his emphasis on witnessing and the display of electrical effects using the orthodox apparatus.

CONCLUSIONS

My aim in this paper has been twofold. First, I have attempted to recover the culture within which the science of electricity was constructed during the 1830s and early 1840s. I have suggested that this culture also provided the evidential context for the first attempts in England and America at constructing viable telegraphic devices. Second, by focusing on Morse's single contribution on telegraphy to a natural philosophical journal I have displayed some of the strategies that could be employed by a telegraph inventor to take his device out of the laboratory and into the world.

In conclusion, I would like to return briefly to the historiographical issues outlined at the start of this paper. What does the story I have given about the telegraph tell us about the general issue of the science–technology distinction and its validity as an analytic category? First, I would suggest, my account implies that any attempt at straightforwardly categorizing Samuel Morse as a technologist and Michael Faraday (for example) as a scientist would do injustice to the social categories employed by the actors in question. At no stage during the early history of the telegraph were distinctions between science and technology made by the actors themselves. On the contrary, telegraphists and electricians formed part of the same cultural group. The telegraph, as Morse's paper demonstrates, was a scientific instrument just as much as it was a means of communication.

Finally, and more interestingly, my paper implies that if there is a use for the science–technology distinction, it should be inverted. It is not the case that technology is simply the routine working out of scientifically established natural facts and application of them to practical ends. On the contrary, the artefacts of what we retrospectively label as technology serve a crucial function in the stabilization of natural facts themselves. A scientific claim such as, for example, Lenz's law was open to debate during the 1830s. No amount of 'scientific' work inside the laboratory could make it a fact. The telegraph succeeded in exemplifying Lenz's law on a large scale and made it an everyday, unproblematically accepted feature of technological *and* scientific practice. This was not an easy task. It required the mobilization of vast resources.[57] Ironically, the conclusion of my paper might well be that Morse—who does not apparently merit an entry in the *Dictionary of Scientific Biography*—owes less to 'science' than 'science' does to Morse and his fellow inventors of the telegraph!

Notes and References

1. Remark attributed to Samuel Morse aboard the packet-ship *Sully*, 1832. See Edward L. Morse, *Samuel F.B. Morse—His Letters and Journals*, 2 vols, (New York, 1914), 2nd vol., 6.

2. Samuel F.B. Morse, 'Experiments made with one hundred pairs of Grove's battery, passing through one hundred and sixty miles of insulated wire', *Silliman's American Journal of Science*, 1843, 45: 390–4.

3. For an account of the 'old' view of the science–technology relationship and an outline of more recent approaches see Barry Barnes, 'The science–technology relationship: a model and a query', *Social Studies of Science*, 1982, 12: 166–72.

4. See George Wise, 'Science and technology', *Osiris*, 1985, 1: 229–46, for a contextual review of some of this literature.

5. For a survey of this new literature and some of its implications see Thomas P. Hughes, 'The seamless web: technology, science, et cetera, et cetera', in Brian Elliot (ed.), *Technology and Social Process*, (Edinburgh, 1988), 9–19.

6. The literature in this field is by now far too large and diversified for any comprehensive listing. The most important recent sources are Harry M. Collins, *Changing Order: Replication and Induction in Scientific Practice*, (London, 1985); Bruno

Latour, *Science in Action: How to Follow Scientists and Engineers through Society*, (Milton Keynes, 1987); and Trevor Pinch, *Confronting Nature: The Sociology of Solar Neutrino Detection* (Dordrecht, 1986).

7. Thomas P. Hughes, *Networks of Power: Electrification in Western Society*, (Baltimore, MD, 1983). For a summary of Hughes's arguments see the review by Barry Barnes, *Social Studies of Science*, 1984, 14: 309–14.

8. See, for example, Michel Callon, 'The state and technical innovation: a case study of the electric vehicle in France', *Research Policy*, 1980, 9: 358–76, and especially the contributions to Michel Callon, John Law and Arie Rip (eds), *Mapping the Dynamics of Science and Technology*, (London, 1986).

9. Trevor Pinch and Wiebe Bijker, 'The social construction of facts and artefacts: or how the sociology of science and the sociology of technology might benefit each other', *Social Studies of Science*, 1984, 14: 399–441. A more recent version of this paper is to be found in Wiebe Bijker, Thomas P. Hughes and Trevor Pinch (eds), *The Social Construction of Technological Systems*, (Cambridge, MA, 1987), 17–50, which also contains many other interesting contributions.

10. For the notion of 'evidential context' see Pinch, op. cit. (6).

11. The concept of the electromagnetic network was developed in David Gooding, 'Experiment and concept formation in electromagnetic science and technology in England in the 1820s', *History and Technology*, 1985, 2: 151–76. See also David Gooding, ' "Magnetic curves" and the magnetic field: experiment and representation in the history of a theory', in David Gooding, Simon Schaffer and Trevor Pinch (eds), *The Uses of Experiment*, (Cambridge, 1989), 183–223.

12. Faraday was Fullerian Professor of Chemistry at the Royal Institution while Daniell was Professor of Chemistry at King's College London.

13. For an account of the organization of the Royal Institution see Morris Berman, *Social Change and Scientific Organization: The Royal Institution 1799–1844*, (London, 1978).

14. See William Sturgeon, 'Improved electro-magnetic apparatus', *Transactions of the Society of Arts*, 1825, 43: 37–52.

15. Quoted in Richard Altick, *The Shows of London* (Cambridge, MA, 1978), 377. Altick provides a fascinating account, which places institutions such as the Adelaide Gallery firmly in the context of the wide range of edifying (and unedifying) entertainments and displays available to the Victorian public.

16. Quoted (undated and uncredited) in W.H.G. Armytage, *A Social History of Engineering*, (Cambridge, MA, 1961), 146.

17. Edward M. Clarke, 'Mr. E.M. Clarke's description of his electrical magnetic machine', *Phil. Mag.*, 1836, 3rd series, 6: 262–6.

18. Ibid., 264.

19. Joseph Saxton, 'Mr. J. Saxton on his magneto-electrical machine, with remarks on Mr. E.M. Clarke's paper in the preceding number', *Phil. Mag.*, 1836, 3rd series, 9: 360–5.

20. For a more exhaustive account of this issue see Iwan Rhys Morus, 'The sociology of sparks: an episode in the history and meaning of electricity', *Social Studies of Science*, 1988, 18: 387–417.

21. Alfred Smee, 'On the galvanic properties of the metallic elementary bodies, with a description of a new chemico-mechanical battery', *Phil. Mag.*, 1840, 3rd series, 16: 315–21.

22. Michael Faraday, *Experimental Researches in Electricity*, (London, 1839), 7th series, paragraph 217.

23. William Sturgeon, 'Fifth memoir on experimental & theoretical researches in electricity, magnetism, &c.', *Annals of Electricity*, 1840, 5: 121–35 and 293–301.

24. Ironically, Faraday has often been characterized as a 'qualitative' scientist. While he was not as concerned with measurement as later electrical scientists and engineers, such as Maxwell, Thomson or Fleeming Jenkin, it was certainly an issue in his work. For Faraday and quantification see Frank A.J.L. James, 'Michael Faraday's first law of electrochemistry' in John T. Stock and M.V. Orna (eds), *Electrochemistry Past and Present*, (Washington, DC, 1989), 32-49.

25. For a more detailed and comprehensive account of these two cultures and their institutional settings see Iwan Rhys Morus, *The Politics of Power: Reform and Regulation in the Work of William Robert Grove*, (Unpublished PhD thesis, University of Cambridge, 1989), 12-64.

26. Auguste de la Rive, 'Some notes on the present state of the study of electricity in England, collected during a recent sojourn in that country', *Walker's Electrical Magazine*, 1845, 1: 100-7.

27. Ibid., 105-6.

28. [Charles V. Walker], 'Mr. Gassiot's electrical soirée', *Walker's Electrical Magazine*, 1845, 1: 537.

29. John P. Gassiot, 'An account of experiments made with a view of ascertaining the possibility of obtaining a spark before the circuit of the Voltaic battery is completed', *Phil. Trans.*, pt 1, 1840: 183-92. The paper was received by the Royal Society on 11 October 1839 and read on 19 December. Gassiot was not at that time a Fellow of the Royal Society but was elected the following year. Jacobi's work on sparks before contact appeared in M.H. Jacobi, 'Sur la vitesse avec laquelle se développe l'électricité de contact dans un simple couple d'éléments', *St Pet. Acad. Sci. Bull.*, 1838, 3: col. 333-5.

30. Gassiot, op. cit. (29), 187.

31. For an account of the role of witnessing in experimental natural philosophy see Steven Shapin, 'Pump and circumstance: Robert Boyle's literary technology', *Social Studies of Science*, 1984, 14: 481-520. Also Steven Shapin and Simon Schaffer, *Leviathan and the Air-Pump: Hobbes, Boyle and the Experimental Life* (Princeton, NJ, 1985), esp. 55-60.

32. For a survey of the institutional setting of American science during the first half of the nineteenth century see Alexandra Oleson and Sanborn C. Brown (eds), *The Pursuit of Knowledge in the Early American Republic: American Scientific and Learned Societies from Colonial Times to the Civil War*, (Baltimore, MD, 1976). Of particular interest is Nathan Reingold, 'Definitions and speculations: the professionalization of science in America in the nineteenth century', 33-69.

33. For Saxton see his entry in *DSB* vol. 12: 131-2. Wheatstone's work on the velocity of electricity is discussed in Brian Bowers, 'Faraday, Wheatstone and electrical engineering', in David Gooding and Frank A.J.L. James (eds), *Faraday Rediscovered: Essays on the Life and Work of Michael Faraday, 1791-1867*, (London, 1985), 163-73.

34. See their biographies in Clark A. Elliott, *Biographical Dictionary of American Science*, (Westport, CT, 1979).

35. Nathan Reingold (ed.), *The Papers of Joseph Henry*, (Washington, DC, 1979-), 5 vols.

36. Joseph Henry, 'On some modifications of the electromagnetic apparatus', *Transactions of the Albany Institute*, 1828, 1: 22-4. Reprinted in *The Scientific Writings of Joseph Henry*, (Washington, DC, 1886), vol. 1, 3-7.

37. Joseph Henry, 'On the application of the principle of the galvanic multiplier to electro-magnetic apparatus, and also to the development of great magnetic power in soft iron, with a small galvanic element', *Silliman's American Journal of Science*, 1831, 19: 400-8. Also Joseph Henry, 'An account of a large electro-magnet,

made for the laboratory of Yale College', *Silliman's American Journal of Science*, 1831, 20: 201–3. Nathan Reingold, 'Joseph Henry', *DSB* vol. 6: 277–81, contains a good account of Henry's early work. He contrasts Henry's experimental interests with those of Michael Faraday and comes to much the same conclusion as is presented here. 'In the Albany experiments he wanted to design devices suitable for class-room demonstration, that is to get large effects from small inputs. Henry's electromagnets exemplified this on a large scale. When he applied them to demonstrate the long-predicted production of electricity from magnetism, the distinction between Henry and Faraday as experimentalists became evident. Faraday devised ingenious experimental setups to detect small effects; Henry, almost anticlimactically, devised procedures for rendering small effects grossly tangible.' 278.

38. W. James King, *The Development of Electrical Technology in the 19th Century: the Telegraph and the Telephone*, (Washington, DC, 1962), 281.

39. For Page see Robert C. Post, *Physics, Patents, and Politics: A Biography of Charles Grafton Page*, (New York, 1976). Post also presents some interesting evidence which suggests that by mid-century, and particularly following Henry's move to the Smithsonian Institution, the situation in America had shifted in a manner which parallels the British scene. By this time a rift was rapidly developing between those, such as Henry, who wished to establish 'science' as an elite occupation and others, such as Page, who had a more 'democratic' and 'practical' vision of the social role of both science and scientists. Henry, in other words, was moving towards a vision of science which identified him more with Michael Faraday than William Sturgeon.

40. For Webster's biographical details see Elliott, op. cit. (34), 269.

41. Charles Grafton Page, 'Method of increasing shocks, and experiments with Prof. Henry's apparatus for obtaining sparks and shocks from the calorimotor', *Silliman's American Journal of Science*, 1836, 31: 137–41.

42. Post, op. cit. (39), 18.

43. Charles Grafton Page, 'The production of galvanic music', *Silliman's American Journal of Science*, 1837, 33: 396–7.

44. See the list of Page's papers in the *Royal Society Catalogue of Scientific Papers*, 1867–79, 4: 735–7.

45. For an early history of the telegraph which notes the extent to which the early telegraphists utilized the whole range of what is described here as the technology of display see George B. Prescott, *History, Theory, and Practice of the Electric Telegraph*, (Boston, MA, 1860).

46. For a description of the instrument see 'Wheatstone and Cooke's electric telegraph', *Mechanics' Magazine*, 1840, 33: 161–70.

47. Interestingly, one of the main factors which aided the Wheatstone and Cooke telegraph's success by bringing it to public attention was its use to capture a murderer who attempted to elude his captors by train. A message was transmitted down the line while the train was in transit between two stations and the murderer apprehended as he left the train at the next station. *London Illustrated News*, 28 November 1846: 339.

48. 'Mr. Bain's new system of electro-telegraphic communication', *Mechanics' Magazine*, 1847, 46: 590–2.

49. Morse, op. cit. (1).

50. Ibid., vol. 1: 236.

51. Morse, op. cit. (2).

52. Morse was presumably referring to the research published in Emil Lenz, 'Ueber das Gesetz der Leitungsfahigkeit für Electricität bei Drathen von verschiedenen Längen und Durchmessen', *St Pet. Acad. Sci. Mem.*, 1838, 3: 187–204.

The paper had originally been presented to the Academy in 1834 and an English translation had appeared in *Taylor's Scientific Memoirs*, 1837, 1: 311–24. Lenz's work on electrical conduction was apparently more well-known in America than that of his fellow-German, Ohm; see for example Jonathan H. Lane, 'On the law of electric conduction in metals', *Silliman's American Journal of Science*, 1846, 2nd series, 1: 230–41. Ohm's work was certainly almost unknown in Britain until well into the 1840s.

53. For an account of the genesis of the Grove cell see Morus, op. cit. (25), 46–64. Grove himself disagreed with the commonly held view of the relative expense of his battery. See [William R. Grove], 'On the progress made in the application of electricity as a motive power', *Literary Gazette*, 1844, 28.

54. For a sociological analysis of the role of such 'surrogates' in scientific practice see Collins, op. cit. (6), 105. Also Morus, op. cit. (20), 406.

55. Samuel F.B. Morse to Joseph Henry, 31 July 1843. Quoted in Reingold, op. cit. (35), vol. 5: 374.

56. Peter Barlow, 'On the laws of electro-magnetic action', *Edinb. Phil. J.*, 1825, 12: 105–14.

57. Several of the actors were themselves aware of the crucial role played by the telegraph. See James Clerk Maxwell, *A Treatise on Electricity and Magnetism*, 2 vols, (Oxford, 1891), 3rd edn, vol. 1: 'The important applications of electro-magnetism to telegraphy have also reacted on pure science . . . by affording to electricians the use of apparatus on a scale which greatly transcends that of any ordinary laboratory', vii.

Electromagnetic Engines

Pre-technology and Development Immediately Following Faraday's Discovery of Electromagnetic Rotations

BRIAN GEE

INTRODUCTION

To understand the position of Faraday in relation to the early evolution of the electric motor we shall first reflect upon the rapid rise of electromagnetism following Oersted's momentous discovery. In 1820 nothing was known about the magnetic effect of an electric current. Indeed, the nature of the current itself was badly understood and despite all attempts to connect magnetism and electricity it was generally concluded that these phenomena were distinct imponderable fluids.[1] Nevertheless, some philosophers had observed that needles would become magnetized during lightning storms and so the conception of a link persisted. Thus Oersted's chance observation, in December 1819, of the deflection of a compass needle on interrupting a nearby voltaic circuit was of more than passing interest.[2] Therefore the so-called discovery of electromagnetism was not entirely serendipitous; realizing its importance Oersted broadcast news of it in Latin rather than his native tongue (Danish) in order to ensure a rapid dissemination. This gave rise to simultaneous investigations in Switzerland, France, Britain and Germany.[3]

By 1826 the initial phase of cognitive effort was finished, largely as a result of Ampère's programme.[4] By June 1822 he had the relevant law of force between two current elements, and by 1826 his theory of electromagnetism was generally accepted by scientists. That is, his masterly synthesis and superbly conducted tests, collectively presented in *Recueil d'observations électrodynamiques* (1822), had become formal theory by the time they reappeared as *Mémoires sur la théorie mathématique des phénomènes électrodynamiques uniquement déduite de l'expérience* (1826). Interpreting these advances in terms of practical electrodynamics was the work of J.F. Démonferrand, whose classic text, the *Manuel d'Électricité* (1823), became of considerable importance in England, particularly following James Cumming's translation of the work in 1827.

Faraday's position in the line of these electromagnetic developments was, at first, of a lesser standing than that of his contemporaries, such as Peter Barlow in Britain and perhaps Jean-Baptiste Biot in France. Nevertheless, as we now understand, it was through his exploration of the subject from scratch, particularly his findings concerning electromagnetic rotation, that the field concept was born. Indeed, it was from the results of this very work that he emerged from Davy's shadow at the Royal Institution to gain the respect of the scientific community at large. Even so, his notions of what was taking place in the space surrounding a current-carrying conductor, as well as within the conductor itself, were, in 1821 and 1822, far from being settled. Certainly he was not in a position to do battle with Ampère's analytical approach, based as it was on the seemingly surer grounds of Newtonian action-at-a-distance.[5]

The various stages of interaction between thought and experiment which culminated in Faraday's now celebrated demonstration of electromagnetic rotations are adequately covered elsewhere.[6] Nevertheless, it is worth restating here that he was led to his discovery partly through scrutiny of Oersted's work, in which he found certain inconsistencies, and partly through doubts about Ampère's material conception of an electric current. Carrying out Oersted's experiments afresh had shown him that there was a tendency for a current-carrying wire to revolve about a magnetic pole. His painstaking approach is charted in his *Diary*, for which the entry of 3 September 1821 reads: 'The effect of a wire is always to pass off at a right angle from the pole, indeed to go in a circle around it.'[7] With this crucial principle in his grasp he was able to construct a crude apparatus to show the circulatory motion of a fly-wire around a magnet. This comprised a vertically mounted magnet, held firm by wax in a deep vessel and surrounded by mercury, leaving the upper pole (effectively a monopole) showing. The suspended fly-wire was connected to a battery via the mercury.[8] This well-known arrangement, together with the additional means of showing the converse effect of a magnet rotating about a fixed wire, was later transformed into a professional demonstration piece by the instrument-maker Newman[9] (Figure 1).

What Faraday had achieved, namely the 'circular force' or, more precisely, the tangential force causing a current-carrying wire or magnet to move in a circle, was altogether a new idea. Naturally it was a challenge to Ampère, whose explanation of electrodynamic action was generally based on the concept of a central force, that is, the interaction between two current elements *along* a joining line in contradistinction to Faraday's force resulting at *right angles* to the joining line. Of course, Ampère took up the challenge and was able to show that the new observation was not discordant with the basic tenets of his theory. Others joined in the theoretical debate but the essential detail, in this context, was the effect itself which quickly became part of the philosopher-mechanics' repertoire. Even without any theoretical understanding, this new 'trick' for making things move was employed to produce a variety of rotational devices.

At this point it should be remembered that—in the parlance of a patent writer—an electric motor is a device for delivering mechanical power by the application of electromagnetism. Furthermore, any definition of a

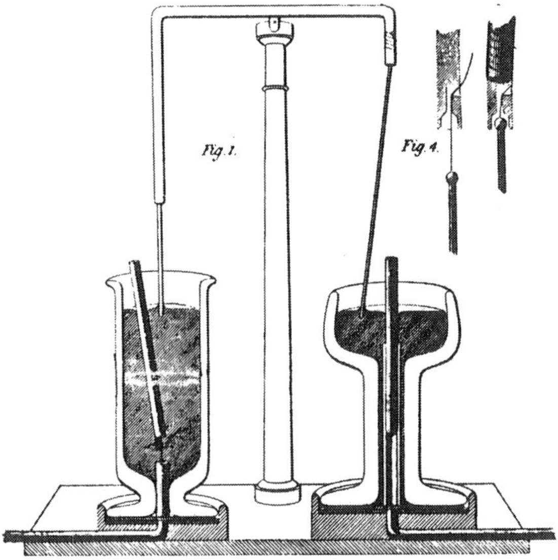

Figure 1 Faraday's apparatus to show the rotary effects of a magnet about a fixed wire and a wire about a fixed magnet. (*By courtesy of the Royal Institution, London.*)

motor must include some account of the fixed and rotating magnets (the stator and rotator) and also some account of the mechanism of current reversal (the commutator). Only with these qualifications can a rotary device be admitted to this category. Thus Faraday's crude arrangement to show the compound rotation of a bar magnet and fly-wire does not constitute an electric motor and, therefore, its invention should not be attributed to him. Nevertheless, the tangible evidence of continuous rotary motion from electromagnetism did inspire others to develop the idea. Inevitably then, attention must shift away from Faraday as 'discoverer' of a new electromagnetic principle to those practitioners whose perseverance and ingenuity in articulating the effect in the engineering sense did eventually give rise to the simple electric motor. That, however, was some ten years into the future.

DRAMATIS PERSONAE

Before presenting specific examples of what constitutes the pre-technology of the electric motor the following account is intended to guide the reader to those individuals, not particularly well known in the annals of the subject, who were responsible for the *mise en scène* in the 1820s.

Peter Barlow

The Royal Institution, where Faraday worked, did not at this time have an official instrument-maker, although John Newman, a local tradesman then at Lisle Street, was sometimes called upon for special projects, such as making up the sturdy version of Faraday's apparatus for electromagnetic rotations. At the Royal Military Academy at Woolwich, however, Peter Barlow was fortunate in having the technical services of James Marsh from the 'shop' at the nearby Royal Arsenal, where he acted as dispensing chemist to the surgeon's apothecary. Also fortunate in this respect was Ampère, at the École Polytechnique in Paris, who employed the services of the younger Pixii from a local instrument *atelier*. In fact the Royal Military Academy (founded 1741) was an institution similar to the École Polytechnique (founded 1795) in its aim to produce military engineers, and the parallel situation, with maker and user working in harmony to produce demonstration apparatus for teaching purposes, was, it seems, an important factor in accounting for the rapid development of electrodynamic apparatus.

Barlow was an extremely competent mathematician and engineer with an important piece of research on the strength of timber already to his credit.[10] He had risen from obscurity, through an assistantship in mathematics, to become a fully fledged professor at the Academy. Well before Ampère had begun his research programme, Barlow was already engaged in the subject of magnetism on behalf of the Board of Longitude, which had provided him with a grant to find a way of reducing the unwanted effects on ships' compasses caused by the increasing use of iron in the fitting-up of ships. The theoretical solution to that problem appeared in his *Essay on Magnetism* (1820) and the practical solution, greatly appreciated by his Admiralty sponsors, came through the design of a compensating iron plate, for which he was awarded a gold medal and prize of fifty guineas by the Society of the Arts.[11]

The fortuitous arrival of the knowledge of electromagnetism from the Continent immediately added a further dimension to his magnetic interests. Like Ampère, Biot and Laplace, Barlow joined in the search for some quantifiable expression which might lead to testable effects of the newly discovered phenomena. His premise was that:

> every particle of the galvanic fluid in the conducting wire acts on every particle of the magnetic fluid in a magnetised needle, with a force varying as the square of the distance; but that the action of the particles of the fluid in the wire, is neither to attract nor repel either poles of a magnetic particle, but a tangential force which has a tendency to place the poles of either fluid at right angles to those of the other; whereby a magnetic particle, supposing it under the influence of the wire only, would always place itself at right angles to the line let fall from its perpendicular to the wire, and to the direction of the wire itself at that point.[12]

'If such a force be admitted,' explained Barlow, 'then all results of reciprocal action may not only be explained but computed and that the results should agree with experiments.'[13] Throughout the spring and summer of 1822 he set about making the necessary tests and by October he had turned the 'important and curious subject' into a sequenced lecture programme incorporating some twenty experiments, each of which was demonstrated by an item of electromagnetic apparatus.[14]

In this project Barlow was lucky to have the services of James Marsh. One example, from the many signs of gratitude expressed by him, will suffice here to underline the importance of the technician's role:

> This young man, to whose ingenuity and industry I am much indebted for the success of my experiments is at present employed in an inferior situation in the laboratory of the Royal Arsenal; but his dexterity as a workman, his practical chemical knowledge, and his regular conduct, are qualifications which render his deserving of a more respectable and profitable occupation.[15]

Through Barlow's continued encouragement Marsh gained due recognition from the Society of Arts in 1823, when he was rewarded with the Society's silver medal and thirty-guinea prize for a portable electromagnetic chest. It is interesting to note that Pixii was at this time in the business of constructing a philosophical demonstration table for electromagnetism for Ampère.[16] Thus in both England and France electromagnetic science was already set for general diffusion.

William Sturgeon

William Sturgeon's position in these important electromagnetic developments at Woolwich followed his demobilization from the Royal Artillery in 1820.[17] At thirty-seven years of age and without any employment he first distanced himself from the multitudes in a similar situation who had returned from the Napoleonic campaigns. It may have seemed a good idea at the time to return to his native Lancashire in order to continue the family shoe-making business, although this was less agreeable to his wife, whose family origins were in Woolwich. Not surprisingly perhaps, he returned to Woolwich and the military establishment, though now as a civilian plying his trade with five other cobblers in the stomping ground at Artillery Place. Although impecunious, he persevered with self-improvement through the Woolwich Literary Society where, in 1822, he came into contact with James Marsh.

Reminiscing on this chapter of his life during his later Manchester years, Sturgeon recalled how the skills of the instrument-maker were added to those of his original trade. He wrote:

> My first step towards constructing apparatus was the purchase of an old lathe and a few tools to work with. My stock of tools, however, was very limited, from want of means to purchase more; so that many rude substitutes had to be brought into requisition. And as I had no practice either at the lathe or the vice, my first essays in apparatus making were necessarily deficient of those elegancies of structure and refinement of workmanship, which always characterise the production of regular-bred instrument makers.

Thus, by dint of these efforts, he became known for his instrument-making exploits. Continuing his autobiography, he related:

> Notwithstanding these disadvantages, however, *hope* and perseverance surmounted every difficulty of constructing such pieces of apparatus as I immediately wanted for my own purpose; and it was not long before I found purchasers for others, even amongst the first-rate makers in London. I am not aware that my progress was due to any particular genius that I possessed, although I never had even a moment's instruction from any person; and I record these particulars with no other view than that of encouraging others to persevere, who may have had a similar disadvantage to encounter.[18]

Indeed, it was but a short step for Sturgeon to become known to Barlow and his colleague S.H. Christie.[19] The link with the Military Academy will have become even firmer in 1824 when Sturgeon began to lecture at the East India Company's Seminary at Addiscombe, for it was from here that cadets, already versed in military subjects and Hindustani, passed to Woolwich for their final training before qualifying for service in India.

From this time Sturgeon began to give serious consideration to constructing his own set of apparatus for lecture illustration. 'The science of electromagnetism,' explained the cobbler turned teacher, 'although so generally interesting, yet (comparatively speaking) appears to be very little understood. The latter circumstance is probably, in a great measure, owing to the difficulty of making the experiments, and the great expense attending to the process.'[20] This was putting the problem mildly. Homing in on the nub of the greater problem he explained:

> Another, and perhaps no less obstacle to the advancement of this interesting science, is, that the experiments being hitherto exhibited on so small a scale, are by no means calculated to illustrate the subject in public lectures; for when the experimenter succeeds even to his wishes (which is not frequently the case), the experiment can only be seen by a very near observer, and the more distant part of the auditory are obliged to take for granted what they hear reported (from the persons who are more favourably situated), of some of the most interesting facts, which they, from their distance, are unable to witness.[21]

Sturgeon was thus inspired to improve upon the demonstration apparatus of Barlow and Marsh.

Until Sturgeon attacked the problem of how best to present electromagnetic phenomena, two design features alone, the 'delicacy' or sensitivity of the apparatus and the battery size, were adjusted to attain optimum effects. Of the two, the most favoured excuse for frequent failure of an experiment during a public performance was 'an insufficiency of galvanic power'.[22] A third design variable, largely ignored on practical grounds, was that of increasing the strength of the magnet rather than that of the battery. By applying the electromagnet principle—using batteries weighing only 1 lb. 5 oz. and needing only 20 square inches of plate area, compared with Marsh's relatively cumbersome battery weighing 12 lb.

and having eight plates each of 1 square foot—Sturgeon developed a comprehensive range of demonstration apparatus which, with the academic backing of Christie and Gregory from the Military Academy, he submitted to the Society of Arts in 1825 (Figure 2).

Francis Watkins

The march of electromagnetic researches and that of encouragement of the working population to study 'useful knowledge' moved step by step in a way now generally forgotten. With the rise of mechanics' institutes it was inevitable that a time would eventually arrive when electromagnetism would form an integral part of customary lecture courses in natural philosophy.[23] Of course, there had been a gradual release of research experience from the time that Oersted's effect was announced: Cumming had demonstrated the effects before the Cambridge Philosophical Society, Faraday before the attenders at the Royal Institution and John Tatum before those at the City Philosophical Society.[24] But it was the outcome of the Woolwich research programme that provided the essential structure for any course. Those attending a special course given by Barlow at the London Institution in 1823 were among the very first to benefit from these developments.[25] Sturgeon's cadets at the East India Company Seminary followed close behind.[26]

Signs that the subject had reached the status of acceptable knowledge appear in the literature. Ampère's paper on the mutual interaction between two current elements received its first British airing through a translation in the Royal Institution's *Quarterly Journal* in 1826. This definitive document of electrodynamic understanding reappeared in the *Philosophical Magazine* and subsequently, in serial form, in the *Mechanics' Magazine*.[27] Important though it was, neither this theoretical paper nor Démonferrand's mathematical commentary, just then available in translation as Cumming's *Manual of Electro-Dynamics* (1827), was likely to cause any significant stir among those mechanics who thrived better on noticeable effects than on mathematical theories.

Nevertheless, signs of a practical following for the subject appear in the activities of instrument-makers. In 1827, John Newman of Regent Street, by this time instrument-maker to the Royal Institution, provided an entirely new section to his catalogue by offering a restricted range of electromagnetic apparatus.[28] Faraday's rotary apparatus could be purchased for £1 10*s*., Ampère's version for 7*s*. 6*d*., an apparatus showing the poles revolving in contrary directions for 9*s*. 6*d*., and Barlow's wheel motion for £1 15*s*. 6*d*. Meanwhile, Francis Watkins at Charing Cross had begun to provide public demonstrations of the rotary apparatus he was then constructing. In November 1827, for example, he showed his apparatus to attenders at the London Mechanics Institute.[29] Mindful of this kind of audience, he wrote a *Popular Sketch of Electro-Magnetism or Electro-Dynamics* (1828), which contained the explanation:

> But as many persons may wish to be acquainted with the subject, who, from not being familiar with mathematical expressions and

(a)

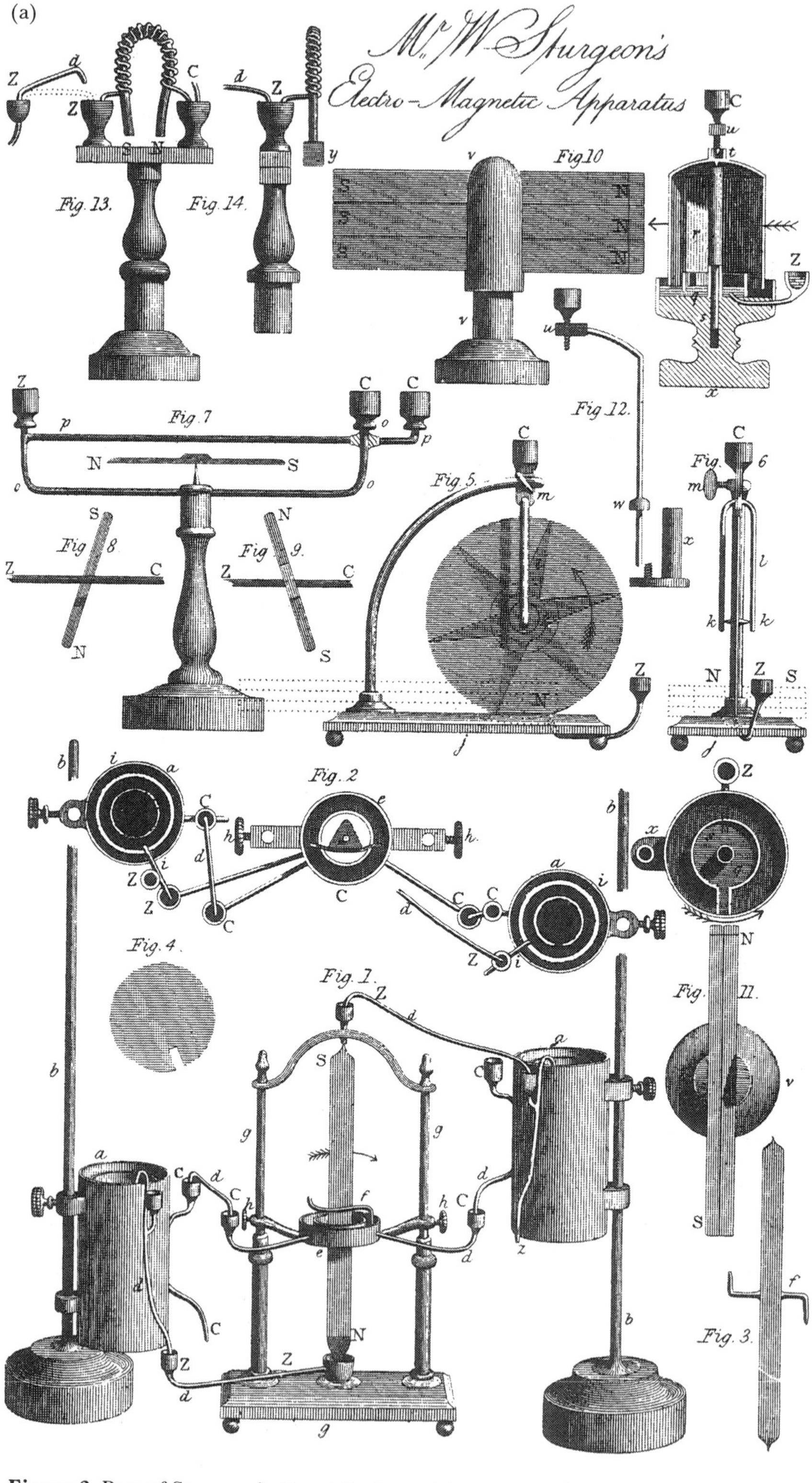

Figure 2 Part of Sturgeon's Royal Society of Arts range of electromagnetic apparatus (1825). (*Figure 2(a) by courtesy of the Deutsches Museum, Munich.*)

(b)

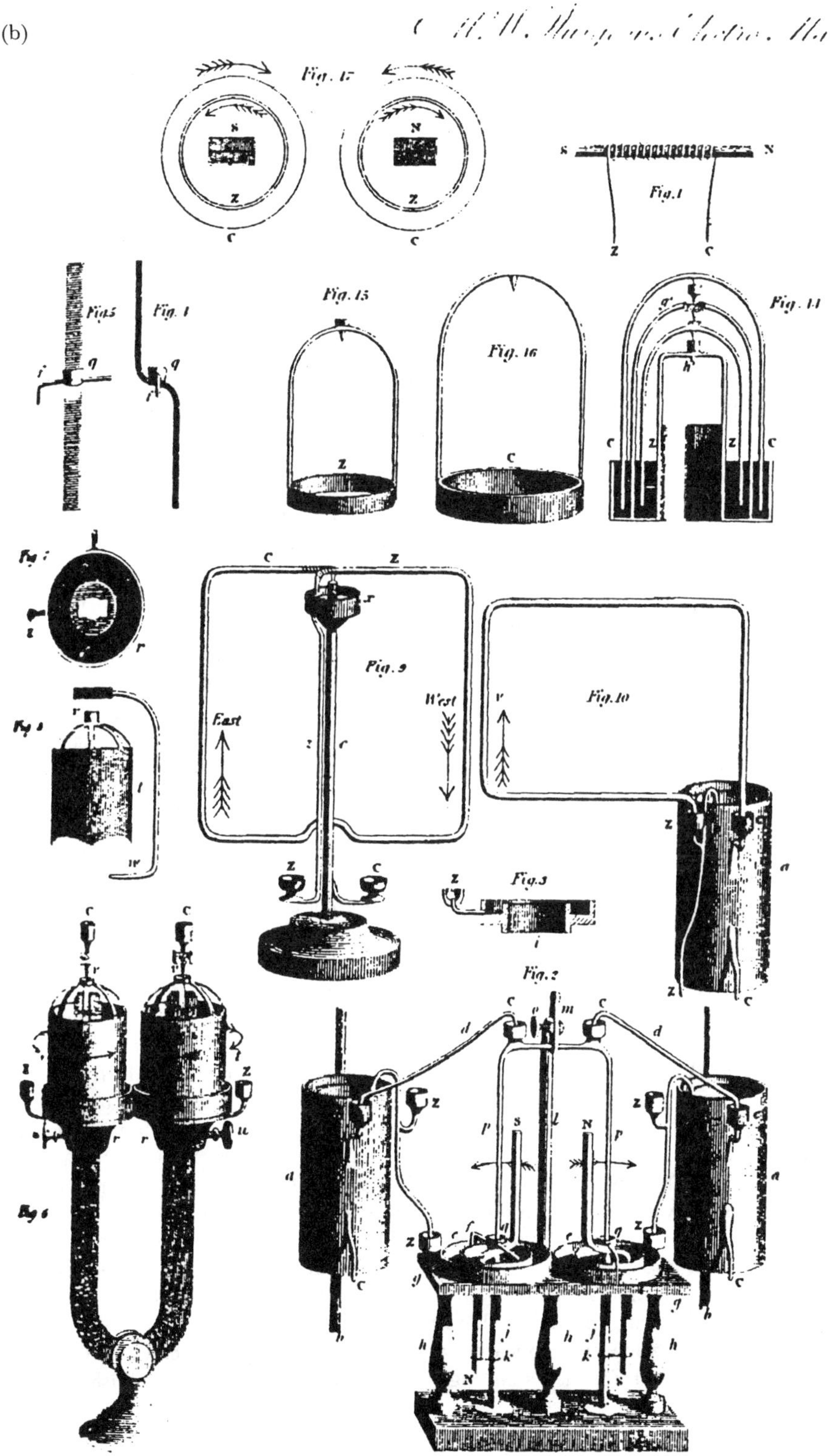

> formulae, would have difficulty in persuing Professor Cumming's work, this *Popular Sketch* has been compiled; and to assist to the same class of inquirers, and in the hope of spreading over a wider field, a more general knowledge of the parent sciences, Electricity and Magnetism—an outline of these sciences has been carefully traced.[30]

Watkins's electromagnetic apparatus in 1828 was much wider in scope than Newman's and by 1832 he had added to the range a 'portable case, containing an Assortment of Electro-magnetical Instruments, which comprises such a series of apparatus, that by proper combinations most of the experimental facts of the science may be illustrated with it.'[31] Thus the developmental stage, begun by Barlow, Marsh and Sturgeon, had now come to commercial fruition and electromagnetic devices were available to those who wished to teach the subject.

The *Popular Sketch* was not a pretentious work although the task of interpreting the history of electricity and magnetism conjointly was a task of some considerable difficulty. As the author unassumingly explained: 'all that is intended, is to lay the facts before the student in such a language as will make the matter clear to him; avoiding all theoretical disquisitions'.[32] For a practitioner like Watkins the 'facts' were the demonstrable truths, the ways and means of making things go. The know-how of demonstration was settled and therefore ready to be released whereas the 'ingenious and plausible' theories were 'far from being satisfactory' and so left unelaborated.

As the first text of its kind, quite unlike Cumming's translation of Démonferrand, it attracted two long reviews.[33] Both reviewers (unnamed) sang praises for the painstaking way in which the instrument-maker had culled the literature: the *Philosophical Magazine*, being somewhat the more generous in its remarks, offered commendation for the author's undertaking to 'make this beautiful and interesting department of knowledge, a branch of *popular* science' while the *Mechanics' Magazine*, more grudgingly, held the effort to be 'unambitious but very meritorious'.[34] Both journals cited the value of the text for experimenters, with its 'clear and connected detail of the principal observations and result of experiment . . . and directions for their performance' cited by the former and the 'many useful hints' cited by the latter. There was also something to be said for the way in which Watkins had treated the history, with the *Philosophical Magazine* seeing merit in the naming of the 'original and first rate authorities . . . instead of only referring to compendiums and compilations', although the *Mechanics' Magazine*, less laudatory again, complained that 'He never quotes his authorities (book, chapter, or verse, we mean) . . . and never makes use of letters of reference.'[35] This last detail would hardly have mattered for an audience of beginners: it was, after all, a work of diffusion, not of scholarship.[36]

Despite the generally favourable reception, no further editions of the *Popular Sketch* appeared at Watkins's instigation even though both reviewers noted corrections and suggestions in anticipation of a reprint.[37] A much more rigorous text offering an impeccable non-historical guide through the maze of electrodynamic apparatus appeared when P.M. Roget, then secretary of the Royal Society and Faraday's colleague at the Royal Institution, published a treatise entitled *Electricity, Galvanism and Electro-Magnetism* (first edition 1827, second edition 1832) for the Society for the Diffusion of Useful Knowledge.[38] This was a work of considerable clarity, containing occasional acknowledgements of Watkins on matters relating to apparatus. By the end of the next decade electricity became a discipline in its own right and a whole generation came to know of the various developments, including that of the electromotive engine and magneto-electric machine, through Henry Noad's summative *Manual of Electricity* (first edition 1839, second edition 1844).

APPARATUS FOR ELECTROMAGNETIC ROTATIONS

A careful examination of the electromagnetic apparatus listed in Watkins's catalogue of 1828 might allow for a division along the following lines.[39] On the one hand there were pivoted compass needles and floating helices capable of alignment along a magnetic meridian, while on the other hand there were rotating devices of various sorts. Items in the former division were designed to show the Oersted effect and the Ampèrian equivalence of a current-carrying coil to a magnet, while those in the latter division were all dependent upon Faraday's principle of electromagnetic rotation. In the present context, it is the items in the latter division, based on the developments and improvements of Barlow, Marsh, Sturgeon and Watkins, which provided the philosopher-mechanic with a vision of what might become a new motor force. These items may be classified under the headings that follow.

The stellar wheel

This famous device may be dealt with briefly. It was a straightforward instrument, said to be due to Barlow, for showing the interactive force between a magnet and movable current-carrying star-shaped disc (replacing the kicking wire). This is shown in Sturgeon's range (Figure 2a). It has survived until recent times (i.e. until the advent of the mercury-free teaching laboratory) because of its simplicity in illustrating the 'motor effect'. The inspiration for this idea was not entirely that of Barlow, but largely that of his technical assistant, James Marsh. In a communication to the *Philosophical Magazine* dated 13 March 1822, Barlow relates:

> After having been repeating Mr. Faraday's rotating experiment the young man who was assisting me wished to try the effect of a horseshoe magnet upon a freely suspended galvanic wire, as it hung with its lower end in the mercury. The wire was immediately thrown into rapid oscillating motion, flying completely out of the mercury; when the contact was broken, it fell by its own gravity to be again projected, and so on, as long as the action of the battery lasts.[40]

Not only did Marsh demonstrate a clear grasp of the principle, he was also able to rejig the effect by replacing the suspended 'kicking wire' with its logical development, the stellar wheel.

Rotating a magnet about its own axis

Faraday's rotation could be expressed in two distinct forms: either the rotation of a magnet about a fixed current-carrying wire, or the rotation of a current-carrying wire about a fixed magnet. Any improvement of the former, beyond Faraday's own version, meant designing an apparatus with greater stability as well as reducing the frictional drag of the magnet through the mercury in order to increase the rotational speed.[41] Barlow did not have such an apparatus in his range although Sturgeon made a double version (with simultaneous clockwise and anticlockwise rotations) for his submission for the Society of Arts prize (Figure 2b). In this apparatus current enters and leaves via the mercury cups and travels through one-half the length of the fixed wire only. Thus the rotation is caused by a moment offered by the pole in the top half of the arrangement.

The advance from the double-bent magnet capable of rotating around an axial current-carrying wire to the arrangement for a straight magnet to rotate around its own longitudinal axis appears to have been made by Ampère, probably with Pixii's assistance.[42] In the version shown in Figure 2a the current enters and leaves via the mercury cups and travels through one-half the length of the fixed magnet only. Thus the rotation is caused by the (unequal) moments offered by the protruding wires.

Rotating a wire about its own axis

Versions of this form of Faraday's rotation usually appeared as the rotation of either a wire cage (regarded as parallel wires) or a cylinder (regarded as an infinite collection of parallel wires) about a single fixed pole. Of all the variants of the rotational apparatus this one proved particularly popular, not least because all visible signs of the power source could be disguised, thereby offering the viewer a mystery, particularly since the cylinder would run for hours without attendance. Causing a cage or cylinder to rotate by a current passing through the arrangement from some external voltaic source required a somewhat complicated design. Essentially the cage or cylinder was mounted on a pivot directly above the pole of a magnet. Ampère's version displayed a certain cleverness by integrating the voltaic cell into the rotating mechanism itself thereby ridding the device of its obvious external connections.[43] Marsh improved upon this by exchanging Ampère's zinc acid-containing cylinder for one of copper because of advantages of relative inertia (the zinc being less massive)—that is, making the zinc cylinder a rotator and the copper cylinder a stator and container for the nitric acid. An additional pivot was then added by Sturgeon to allow both zinc and copper cylinders to rotate simultaneously and oppositely (Figure 2b).[44] Just such a nicety was exhibited by him at the Holborn instrument-makers W. & S. Jones in autumn 1823, according to the 'Intelligence' of the *Philosophical Magazine*, which said: 'Mr. Sturgeon, a pensioned artilleryman of Woolwich who has successfully devoted himself to scientific pursuits, has constructed the apparatus with two sets of rotating cylinders on a horseshoe magnet. The effect is the most pleasing we have ever seen, and was witnessed at the house of Messrs. Jones, optician, Holborn.'[45] Subsequently this object was sold to someone in the United States.[46]

Watkins made rotating cylinders with minor improvements. Whereas Marsh had required an external loop to complete the circuit, Watkins, like Ampère, integrated this, thus making the apparatus self-contained. In other words motion began immediately the acid was poured into the trough and continued until the battery was exhausted. Thus the mystery was complete and unthinking visitors to the Watkins and Hill establishment may well have believed this to be some kind of perpetual motion machine, although, of course, the laws of thermodynamics (though not formally stated at this time) were not confounded. Spectators, if patient, would have eventually seen the acid being replenished and the zinc being replaced.

Thermo-electromagnetic rotations

Another form of rotating cage, alternatively powered by thermo-electricity, was developed by Marsh after he had been given a task to perform by Barlow, based on some recent researches by James Cumming at Cambridge. Cumming had been interested in the possibility of using the weak current from a platinum–silver thermocouple as a sensitivity test for a newly designed galvanoscope following Seebeck's discovery of the thermo-electric effect. Marsh not only copied successfully Cumming's plan for Barlow but also, of his own accord, invented a device which caused 'compound wires' (thermocouples), pivoted on a fine steel point on top of the agate above the poles of a horseshoe magnet, to spin at up to 30 r.p.m., thus offering a 'pleasing exhibition' (see Figure 3a).[47]

It was not long before Sturgeon investigated the possibilities of thermo-electricity. Notwithstanding certain technical difficulties, he concluded: 'my anticipations were soon agreeably realised by witnessing the first thermo-rotation ever produced by the influence of a central magnet'.[48] His arrangement (illustrated in Figure 3b) was driven by a circle of spirit lamps on a stage and said to have reached the 'acme of perfection' when heated by a circular flame of ignited hydrogen controlled by a stop-cock.[49] Unlike Marsh, however, Sturgeon saw achieving the effect not as an end in itself, but only as a beginning for his philosophical speculations on the rotation of the earth.

Through the 'wire machine', as he called it, Sturgeon began to perceive similarities 'imitative' of Halley's (defunct) hypothesis of a hollow earth containing a giant magnet. Of course the model, as it stood, contained only one pole, although it was not long before he contrived both poles to be enveloped within a globe-like cage of galvanized wire. At first Sturgeon was cautious, arguing that 'It is by no means intended from this experiment to assert that the rotary motion of the earth and planets is really the effect of electromagnetism.' Nevertheless, he later came to believe that the earth's rotation was indeed caused by a combination of earth 'magnet' and thermo-electric effect due to a temperature differential between the poles and the equator.[50]

Rotating helices

All the possibilities of rotating devices appeared to be exhausted by the time Sturgeon had offered his range to the Society of Arts in 1825, although it is

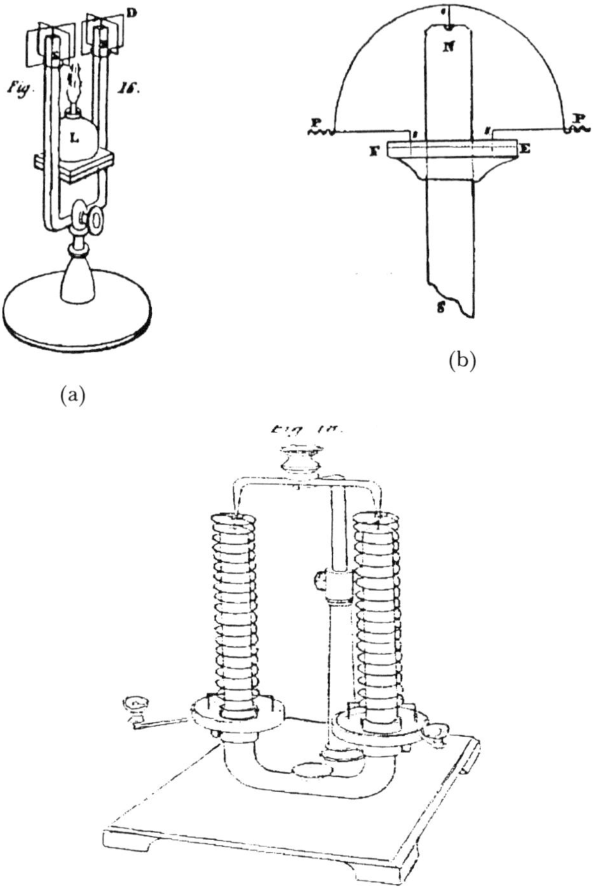

(a) (b) (c)

Figure 3 Other rotary apparatus: (a) Marsh's device; (b) Sturgeon's arrangement; (c) Watkins's rotating helices.

not surprising that the ingenious Watkins should have added his own peculiar contribution by way of a rotating helix (Figure 3c). Introducing the variant into the evolutionary line, he observed:

> Several methods have been proposed for exhibiting the revolution of an electrified wire about the poles of a magnet; but the author flatters himself that that which he has contrived, affords a better example than any that has been yet shown, and certainly has a more pleasing effect. It consists of helical coils instead of straight wires formally used; these wires are suspended on the poles of a horseshoe formed magnet; and when the current is passed along them, they revolve on their axes in the contrary directions. The direction of rotation is reversed, when the direction of current is reversed.[51]

If by 'better example' Watkins meant a simple and more obvious illustration of Faraday's principle then there may be some doubt when his device is compared with, say, the Barlow wheel, but to afford a 'more pleasing effect' was simply to announce a novel form. The *Mechanics' Magazine* reviewer of Watkins's book, however, concurred with the maker by reporting that the article 'well deserves the preference which he claims for it;—being at once elegant, simple and efficient'.[52]

Elegance, simplicity and efficiency are indeed some of the aesthetic and semi-objective criteria by which technological artefacts may be judged. At this stage none of the items described above was intended to perform useful work. Nevertheless, they are state-of-the-art representations of what could be done with Faraday's principle of a tangential force caused by a current-carrying conductor. Before the electromotive engine could fully emerge new ideas were needed which not only demanded redesign of the established rotary philosophical toys but, in fact, involved an entirely new conception in design.

THE FIRST ELECTROMOTIVE ENGINES

Those early rotary arrangements of the 1820s—the Barlow wheel, the Sturgeon disc, the Ampère-Marsh cylinder and the Watkins rotating helices—could never be claimed to be engines because of their inability to deliver useful work. Not until a new design involving the interaction between a permanent magnet and an electromagnet (effectively the field magnet and the moving electromagnetic armature), together with some technique for reversing current in order to provide the mechanical force at the right moment (effectively a commutator), had been produced could the preliminary stage of a technology of (direct current) electric motors begin. These ideas did not occur simultaneously and their order of appearance is not without a certain logic. First, without any obvious foresight of an assured future in electrical power machines, the load-bearing capacity of the electromagnet was increased by trial and error; this resulted in some gain in understanding of the relation between optimum strength and the battery source. Second, out of these experiments grew a familiarity with the retentive property of soft iron and a realization that polarity could be changed with considerable rapidity. Third, in order to utilize this last property in some self-acting engine it became a logical necessity that a contrivance should be arranged for changing the direction of current to the electromagnet. These were the essential principles—first brought together by the American experimental philosopher, Joseph Henry, in 1831 and then reproduced by others in the 1830s—which opened up a modelling stage for the electromotive engine.

Henry's reciprocating engine

The time lapse between Sturgeon's presentation of the electromagnet and a pick-up of interest in it, that is, the separation between the invention and its subsequent development, is demonstrated in the following passage, from 1832, in which Sturgeon wrote:

It does not appear that any very extensive experiments were attempted to improve the lifting powers of the electro-magnet from the time that my experiments were published in the Transactions of the Society of Arts etc. for 1825, till the latter part of 1828. Mr. Watkins, Philosophical Instrument-Maker, Charing Cross, had, however, made them of a much larger size than any which I employed; but I am not aware to what extent he pursued the experiment.[53]

Retrospectively, the year 1828 seems to have been significant for Sturgeon because it marked the beginning of Gerrit Moll's interests in this field.[54] This Dutch experimental philosopher had just then visited London and, it appears, became inspired on seeing an electromagnet at Watkins's shop that could support a nine-pound load.[55] On returning to Utrecht he improved upon this using the soft iron horseshoe which he had expressly purchased from Watkins during his London visit. Later he supported a load of 154 lb. with a magnet, itself weighing 26 lb., excited by a battery containing 11 square feet of zinc.[56] From his descriptions he appears to have been truly amazed at the ability of electromagnets to be multiplied in strength.

The appearance of Moll's account, reprinted in the *American Journal of Science*, caused Joseph Henry, then at the Albany Academy, New York, to examine his claim.[57] By adding turns to an electromagnet he found that there was some limit to the generation of strength, beyond which a second coil was needed in parallel with the first; he also investigated how a battery employing more pairs of plates could increase the effect. Hence he distinguished the 'quantity electromagnet' (using coils in parallel with a Hare-type battery in which the area of the plates was large) from the 'intensity electromagnet' (which responded best to an intensity battery of the Cruikshank type in which there were a large number of plates).[58] His own quantity electromagnet, with an iron core weighing 21 lb. and employing a battery of 72 square inches of zinc, held a load of 750 lb. This strength was improved even further at Yale University, where experiments with a soft iron core weighing 59 lb., wound with 800 feet of wire in twenty-six coils and supplied with current from a battery of 5 square feet plate area, could support 2,000 lb. (nearly 1 tonne!).[59]

The outcome of these experiments gave Henry an intimacy with the retentive properties of soft iron which led to his realization that the polarities of soft iron bars could be alternated rapidly with small currents. From this notion came the very first self-acting electromotive engine capable of further development. Yet Henry wrote of it:

> Not very much importance however, is attached to the invention, since the article, in its present state, can only be considered as a philosophical toy; although in the progress of discovery and invention, it is not impossible that the same principle, or some modification of it on a more extended scale, may hereafter be applied to some useful purpose.[60]

Thus the principles of the modern electric motor were first brought together, although the device remained undeveloped in Henry's hands.

Figure 4a shows Henry's reciprocating engine with its electromagnetic beam (as armature), supported horizontally on an axis through its centre of gravity, and two permanent steel magnets (as field magnets) arranged vertically with their north poles uppermost so as to face the poles of the electromagnet. By a timely alteration in the magnetic polarity, effected by changing the direction of the current inducing the polarity during the cycle, the beam was alternately attracted between the pole pieces of the facing permanent magnet and electromagnet.

The business of translating the reciprocal motion into rotary motion was the work of T. Edmondson of Baltimore in 1834. His arrangement comprised a number of soft iron armatures (*a, a, a, a*) mounted upon radial arms affixed to a horzontal shaft (shown in Figure 4b).[61] These are successively brought into the range of attraction of a stationary electromagnet (*b*). A commutator comprising an elastic contact spring or brush, pressing against metallic segments fixed upon the revolving shaft (*d, d*) interrupts the current long enough to allow each armature to pass out of the magnetic field. Ingenious though this was, it was too delicate to deliver power.

The Davenport engine

No sooner had Henry's Philadelphian friend, Jacob Green, professor of chemistry at the Jefferson Memorial College, witnessed the invention than he offered suggestions for its improvement. Henry, however, declined to spend time upon what was intended merely as an arrangement 'to exhibit the motion'. Like Faraday, he was satisfied that the little engine could answer the purpose for which it was intended, namely to demonstrate the principles. Nevertheless, once the idea was published, he was beleaguered with correspondence from zealous inventors and cranks whose fertile—though not always well-informed—imaginations had been fired with new inspiration; some of them naïvely searched afresh for the perpetual motion machine.[62] Explaining the inundation to Benjamin Silliman some time later, in a letter dated 10 September 1835, Henry complained:

> I have been pestered with letters from almost every state in the Union since my first paper on Electromagnetism appeared in the Journal [*Silliman's*], containing descriptions and plans of machines to be moved by the magnetic power. I have uniformly referred them to the description of my little machine in the Journal and stated that I freely renounced all right to the invention as I consider the machine in the present state of the science only a philosophical toy. It is surprising how many times this machine has been reinvented and described with slight modifications since my first description of it. One invention of this kind was made in Germany another in France and in the last No. of the Edinburgh Journal [Jameson's] but one, a detailed account is given of one by a person in Milan almost precisely the same as mine except, that it vibrates like a pendulum instead of moving like a working beam of a steam engine. The plan of the Brandon Blacksmith, is however the best although not the

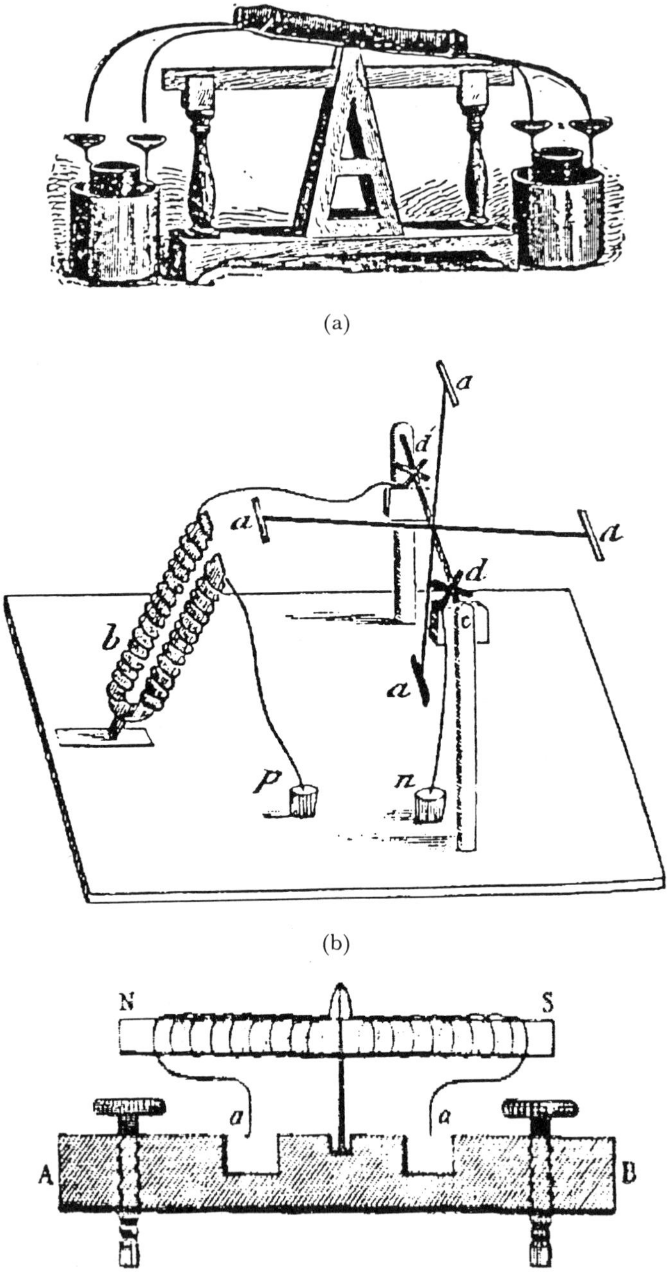

Figure 4 The first electromotive engines (1831–3): (a) Henry's reciprocating beam; (b) Edmondson's engine; (c) Ritchie's engine.

> most simple which I have seen. It is still however nothing more than a philosophical toy—a new power cannot be introduced as a moving principle in mechanics unless it be cheaper in its production, or more convenient in some respect other than steam.[63]

The reference to 'the Brandon Blacksmith' in the above letter was to Thomas Davenport of Forestdale, Vermont, who, according to Benjamin Silliman, had been the first to employ 'make and break' principles to produce rotary motion.[64] He had first been impressed with the possibilities of the electromagnet on seeing the remarkable magnet at the Penfold Iron Works at Crown Point (Lake Champlain), New York, which was used to extract iron from pulverized ore. Instead of purchasing iron, for which he had made the visit, he bought the magnet itself! With the assistance of Orange A. Smalley, he constructed an engine which drove a 7-inch wheel at 30 r.p.m. This was taken to the professors at Middleburg College, Vermont, in December 1834 in search of their approval.

Armed with a letter from his patron, Van Renssaler, Davenport then visited Princeton in the hope of obtaining some certificate of worthiness from Henry. But if Davenport had aspired to fortune as well as fame he was to be put in his place by Henry's scepticism on the economics of such a venture. Nevertheless, the Princeton professor, presumably impressed by the younger man's ingenuity, obliged him by writing a letter of introduction to A.D. Bache, president of the Franklin Institute in Philadelphia, whereupon the engine was exhibited there in Peale's museum (the Franklin Institute Museum) in 1835.[65] Silliman supported Davenport too although, unlike Henry, he saw no reason why the scale and hence power of such machines might not be increased indefinitely. As time went by, Henry was to treat this opinion with some contempt, especially after Davenport's successful application for a patent had received Silliman's personal backing. In consequence, Henry's image as originator of the electromotive engine was dimmed. However, his position was somewhat restored when Sturgeon and Watkins, who had both met Henry during his European tour in 1837, separately acknowledged him as inventor, completely disregarding Davenport.[66]

Sturgeon's engines

The confluence of principles, as outlined above, led to a welter of designs worldwide.[67] At University College, London, William Ritchie was engaged in such a project in 1832–3.[68] His device consisted of a bar electromagnet, mounted so as to revolve freely on an upright axis, its poles at each half-revolution passing close to the poles of two upright permanent bar magnets (Figure 4c). Polarity was reversed during each half-revolution by means of a mercurial commutator. With this device Ritchie managed to raise several ounces via a pulley. An improvement upon his engine, by Charles G. Page of Salem, Massachusetts, in 1838, occurred when a metallic commutator was substituted for the mercury device.[69] Thereafter this engine became a popular teaching aid on both sides of the Atlantic.

It was William Sturgeon, somewhat tardily in 1837, who laid claim to being the first in England actually to power model machinery by an engine of this sort. In an attempt to put the record straight, he wrote:

> This engine [his first] was constructed in the Autumn of 1832, and was exhibited for the first time in London, on the 21st of March, 1833, in a lecture on electro-magnetism which I delivered at the Western Literary and Scientific Institution. And notwithstanding its then rude appearance, the Committee were so highly pleased with its structure and performance, that they expressed a wish to have it brought forward again, and hear it explained as soon as there was another opportunity. I was consequently honoured with an engagement to continue and extend my course of lectures into the following June; and in those lectures my engine again worked well, and excited a great deal of curiosity among the members of the Institution; and I believe was so fortunate as to give general satisfaction.[70]

Those attending Sturgeon's lectures were among the first in London to witness electrically powered models 'for drawing water, waggons, and carriages on a railway, for sawing wood, pumping water, etc.'.[71] Sturgeon had claimed his engines were in proportion to machinery normally employed in conjunction with steam engines, although the engineering problem of how to scale up these miniature engines into those of useful dimensions was not elaborated.

One of Sturgeon's later engines is illustrated in Figure 5a. This comprised a vertical shaft, mounted on a baseboard (*A*, *A*, *A*, *A*), pillars (*B*, *B*) and cross-piece (*C*, *C*). On this shaft are fixed parallel but oppositely mounted magnets (*N*–*S* and *N′*–*S′*). There were four vertically mounted electromagnets (*i*, *i*, *i*, *i*), each comprising six separately wound coils. Appropriate connections were made from these to the four quadrantal vanes mounted on the cross-piece (*D*, *D*) and to the circular mercury channels (not shown), which also contained the necessary intricacies of metallic strips, thereby providing a commutator arrangement for current reversal. Rotation of the shaft occurred by alternating repulsion and attraction between the poles of corresponding electromagnets and permanent magnets when current was supplied from two voltaic cells (not shown). Power was tapped via the geared pulley wheel (*W*). A replica of this engine is illustrated in Figure 6.

Watkins's engines

Two years later, in March 1839, Sturgeon provided an historical review on the progress of 'engines for propelling machinery', in which he related quite precisely when model electrical engines first became available through the trade. He wrote:

> About the latter part of the year 1837 a variety of ingenious contrivances, in the form of rotating engines, were to be seen in the different philosophical instrument makers' shops in London. Some

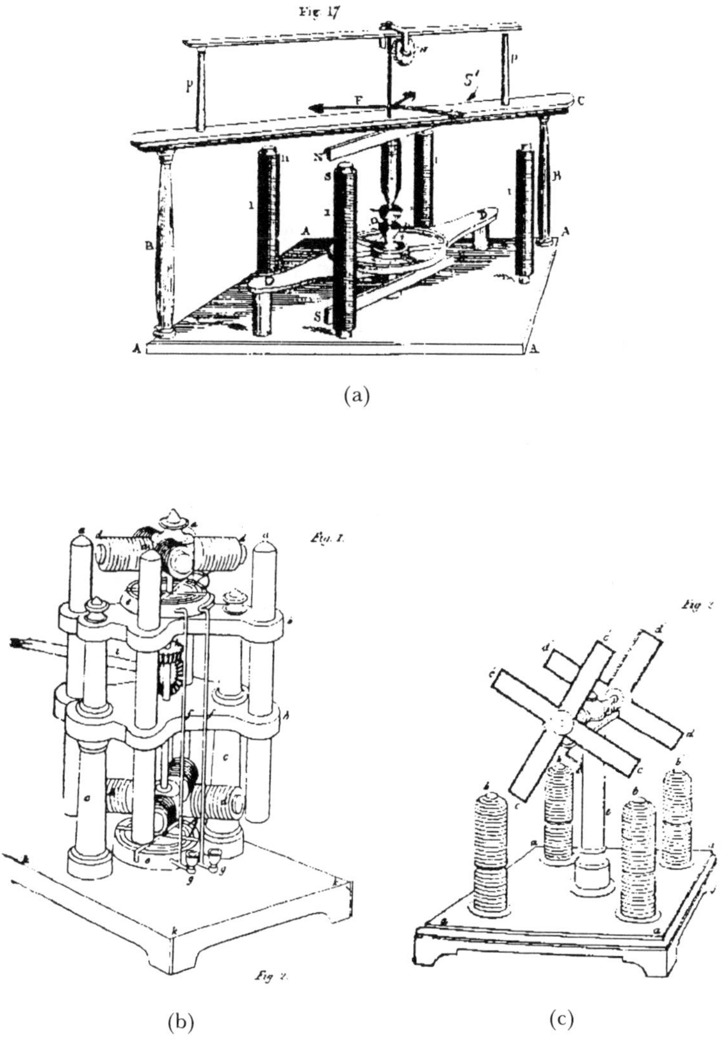

(a)

(b) (c)

Figure 5 Later electromotive engines (1835–7): (a) Sturgeon's engine; (b) Watkins's engine—I; (c) Watkins's engine—II.

of them had miniature saw mills, pumps, and other pieces of machinery attached, and put in motion by them. Mr. Watkins gave a description of one in the Phil. Mag. for February 1838, consisting of four vertical steel magnets, placed at equal distances from each other in the circumference of a circle; and four small iron electromagnets which were fixed at right angles to, and rotated with, a vertical spindle, situated near the axis of the system. Mr. Palmer, of Newgate Street, has long had in his window a very great variety of

Figure 6 Replica of Sturgeon's engine. (*By courtesy of the Science Museum, London.*)

> rotating engines, with attached models of machinery which are kept in motion by them. It is highly probable that every individual instrument maker has some peculiarity in the mode of fitting up this class of apparatus.[72]

In fact, according to the 1838 Watkins and Hill instrument catalogue, 'Watkins's Electro-Motive Machines', offering the viewer 'a fair notion . . . of the probability of such power being economically employed on a large scale by the mechanician', were sold for two, three and ten guineas, for which the purchaser was assured 'a pleasing exhibition on the lecture table'.[73] In 1840 Edward Palmer advertised engines with eight, sixteen or even twenty-four magnets, which sold for prices between five and twenty guineas.[74] Little is known of Palmer's interests or background other than that he had not long been established at Newgate Street.

Watkins had been occupied with the possibility of achieving power-driven motion by electricity for some time, possibly since 1827, when he first began demonstrating the simple rotary apparatus. Through 1832 and 1833 he explored the 'magnetic powers of soft iron', showing that electrically induced magnetism in loaded horseshoe magnets could be retained until the magnetic 'circuit' was broken. As a privileged guest at one of the Duke of Sussex's (then President of the Royal Society) 'conversation parties' at Kensington Palace, Watkins was able to demonstrate this effect by removing the keeper from some electromagnetic horseshoes, which had been pre-loaded some six months previously, whereupon the magnetic effect was completely lost.[75] In 1833 his newly appointed Irish assistant, E.M. Clarke, was assigned the task of devising a means for producing 'locomotion by electricity', although little became of this owing to Clarke's speedy departure to set up business on his own account. Watkins was still engaged in experimental research in 1835 when, at Charles Babbage's instigation, he set out to test Plateau's hypothesis that time was a necessary factor in magnetic action.[76]

The two distinct Watkins engines commercially available in 1838 are illustrated in Figure 5.[77] The version in Figure 5b comprised four vertically mounted permanent steel magnets (*a*, *a*, *a*, *a*) which were held in position by two mahogany cross-stages (*b*, *b*) which themselves were held rigidly by two vertical mahogany columns (*c*, *c*). Two wooden cisterns (*e*, *e*), each containing concentric mercury troughs and each divided into four parts, were connected to a supply via wires (*f*, *f*) and cups (*g*, *g*). The necessary current reversal and consequent polarity reversal in the soft iron were achieved by platinum wires from the coils trailing in the troughs. The lower system of magnets was at an angle with respect to the upper set. Thus one set could function while the other was at some 'dead point'. Power was delivered via a horizontal shaft, connected to the vertical shaft by bevelled gears. Watkins used this with model tilt hammers, pumps and dredgers.

Watkins's other model, shown in Figure 5c, is superficially much simpler. This too was built on a mahogany stand (*a*, *a*, *a*, *a*) on which were mounted a pair of soft iron electromagnets (*b*, *b*, *b'*, *b'*); their cross-pieces were beneath the board and therefore not visible. Four flat permanent steel magnets (*c*, *c*, *c'*, *c'*) and (*d*, *d*, *d'*, *d'*) were arranged in windmill fashion and attached to a horizontal movable axis which itself was supported by a wooden column (*e*). Not clear in this illustration was the arrangement by which platinum discs made contact with mercury in a trough in order to effect the current reversal via wires passing down the central wooden column to the electromagnets. Two pulleys attached to this arrangement were said to 'urge trifling pieces of machinery'.

FEASIBILITY AND IMPROVEMENT

The plethora of invention during the 1830s caused the more cautious to raise questions about the economy of what otherwise appeared to be a wonderful thing. Henry had imposed one criterion by reminding prospective inventors of the futility of competition unless acid and zinc could be made much cheaper than coal and steam. But this neither subdued their enthusiasm nor curtailed their quest for invention and improvement. In any case, there emerged another view in which economic benefit was construed to mean temporal 'convenience', that is, having power available as and when required. It was argued that men, paid by the day, were not expected to be idle, but that having a machine 'on tap' had distinct economic possibilities for those jobs which did not need power of the magnitude of that of the familiar steam horses. A speculative editorial, prepared along these lines for the Franklin Institute *Journal*, caught Sturgeon's attention and was duly reprinted in the *Annals of Electricity*. Certainly the insight contained in this paper is worthy of reproduction yet again (see the Appendix). Similar economic considerations were very much in mind when the young James P. Joule set out to measure the constancy of equivalence between electrical and mechanical powers even if, cost for cost, they were then very far from equivalent.[78]

Watkins displayed an engineer's concern about economy and feasibility when some brash projects were mooted for transforming the small-scale engines into large-scale utilities. 'I have mentioned that many trifling machines or philosophical toys . . . have been constructed,' wrote the emerging electrical engineer, 'and plenty more, I have no doubt, will be brought forward and work successfully.' But, he continued:

> When we reflect that magnetic attractive force is the fundamental principle upon which the motive machines act, the limited space through which this force operates to a working amount, and our imperfect means of developing its powers, it may be excusable if we pause before giving in the present state of our knowledge an unreserved assent to the ultimate success of employing its agency as a prime mover on an extensive scale.[79]

This remark told of the need for a significant improvement in precision engineering as well as warning that the 'fundamental principle' was not yet at hand. Of course, Watkins had no difficulty in envisaging what was needed by way of tooling up for *accuracy*, *stability* and *arrangement*, but he knew that engineers would have to await a further advance from science before design improvements could be considered seriously.

Experiments over the years had tended to suggest that the rapidity with which polarity could be switched in an electromagnet lay within some vague order of magnitude 'many hundred times in a minute'.[80] Implicit in this was a complex design problem demanding some yet unknown calculation that would indicate how best to arrange the permanent and electromagnetic poles so that their optimum (interactive) force might be justly timed. As matters stood, Watkins suspected that maximum rotary speeds had not yet been achieved. As for explaining how the high speeds were maintained, he was restricted to a semiquantitative account in terms of relative inertia and friction. Thus, in testing his own engines he showed that a loaded rotating shaft (that is, one with increased friction in its bearings) resulted in a clear diminution of rotational speed; hence, he concluded that 'under such circumstances we have only the primitive magnetic attractive force of the machine left for mechanical purposes.'[81] Regardless of what, today, might appear as ill-articulated rotational dynamics, Watkins had correctly deduced, demonstrated and explained how the lack of torque in the prevailing model engines had to be overcome by some carefully evaluated scheme of electromagnetic understanding.

Was this hesitancy a sufficient reason to abandon all hope for large-scale electric motors? It was, of course, always conceivable that an improvement could be made by increasing the number of fixed and movable magnets. This much Watkins had appreciated. Advising the aspiring electricians that a little faith was required in the engineering process, Watkins wrote:

> I am well aware it frequently occurs in the application of a philosophical principle or mechanical arrangement that there is a considerable difference between a model and that of a large working model; it therefore behoves all persons experimentally engaged in the application of a principle or a power to bear this in mind, and not to decide too hastily because they fail several times with models.[82]

At the end of this cautionary assessment of the state of the art, Watkins concluded: 'we must look forward and hope for a better knowledge of the nature of the mysterious and invisible agent which is to actuate our machines before complete success crown our endeavours.'[83] These are not the words of a latter-day philosophical instrument-maker but those of a pioneer electrical engineer, for whom the expression 'our endeavours' appears as a goal quite apart from that which occupied the community of experimental philosophers.

Needless to say, any scaling up of the model engines required a massive investment, certainly one that Watkins could not afford without suitable backing. The instrument-maker knew that patronage in Britain was a hopeless quest, and that he was certainly unlikely to attract Treasury funds, particularly after the government's recent experience in backing Charles Babbage's dubious exploits to manufacture a calculating machine.[84] But a new age had dawned in which technological ideas and engineering projects demanded investment on an unprecedented scale that was quite out of proportion with anything he (or indeed a government) might wish to afford. Watkins therefore remained content to observe, with an intelligent eye, events as they occurred elsewhere. Summing up the predicament in 1838, he wrote:

> Human perseverance has achieved wonders; and as the subject engrosses considerable attention just now, and we rejoice to find by the periodicals that the Emperor of Russia has placed at the disposal of M. Jacobi and a scientific committee £500 for the purpose of making experiments, we may indulge in the hope that before long some successful results will be the fruits of their labours, and that a new method of employing magnetism will be discovered; for from the present mode, if my notions are correct, we have little to hope for on a large scale, and playthings are not worth the mechanician's notice.[85]

Watkins showed optimism tempered with a good measure of caution. This was not the attitude of the future navigator of the Neva, the Dorpat civil engineer Moritz von Jacobi, who had built an electric engine that could lift 10 lb. at a speed of 1 foot per second. But he too said that these engines were little more than 'an amusing plaything with which physical cabinets deserve to be enriched'. Even so, he showed that half a pound of zinc would provide the 'demi-force d'un homme' for 8 hours, and generally concluded that 'the superiority of this new mover is placed beyond a doubt, as regards the absence from all danger, the simplicity of application, and the expense of attending to it.'[86] By 1837 Jacobi had succeeded in relaying his convictions to the Tsar, who granted the means by which to prove that his latest electric motor had application. In return the Tsar was to be impressed with an electric motor-boat, powered by 320 Daniell cells, which was later claimed to have speeded a dozen Russian officials along the Neva at one-and-a-half miles per hour.[87] This piece of publicity, or extravaganza of Imperial amusement, may have gained Jacobi his election to the St Petersburg Academy of Sciences in that same year, but equally, for Watkins, it must have confirmed the justness of his cautious attitude, for the expensive electric boat offered no serious competition to a man-propelled craft. The feeling was summed up succinctly by another London practitioner, William Leithead, who had met Henry on his European tour. Relaying the scientific gossip to Henry in October 1839, he wrote: 'Jacobi has a ten-oared boat; but by electro-magnetism it moveth not!'[88]

Others pursued the improvement of the electromotive engine from this point. The Foster Lane instrument-maker George Knight Jr was among those who kept up the pace. He entered a version of his own design for the second British Association exhibition held in Birmingham in 1839.[89] In the USA, Charles G. Page, who held a chair of chemistry at Columbian College, advanced the design of the reciprocating engine in the 1840s. Impressed by his ardour Congress voted him $50,000 to construct an electric locomotive, which ultimately had some modest success in April 1851 when, powered by 100 Grove cells, it drove an engine along the Baltimore–Washington railroad, covering some five miles in 39 minutes.[90] Even so, the prospects for a good powerful engine were considered remote by the jurors of the London Great Exhibition, who stressed that this objective still claimed the attention of 'scientific men'. Watkins's caution was proved wise over and over again and had he lived to read the jurors' report he might have smiled knowingly at their summing up, which said: 'although no great power has yet been obtained [from these machines], many important difficulties in its practical application seem to have been overcome, particularly by Mr Hjorth [of Denmark], and we cannot help flattering ourselves that the attainment of this mysterious motive force will soon be followed by the making it available for practical purposes'.[91]

Thus the engineers' dream of powered rotary motion by electricity was already realized by the mid-century, although the technology was not to take off in any spectacular way until the last decades of the century.

APPENDIX: SPECULATIONS RESPECTING ELECTRO-MAGNETIC PROPELLING MACHINERY BY THE EDITOR OF THE *JOURNAL OF THE FRANKLIN INSTITUTE*[92]

In our number for November last, Vol. XX, p. 340, we published the specification of Mr. Davenport's patent for a machine intended to furnish a motive power by the agency of Electro Magnetism, to which we append some remarks upon the subject generally. We had hoped, ere now, to have received more definitive information than has transpired respecting the progress of the experiments which are being made in New York with a view to its testing the utility by applying it to drive a Napier Press, requiring a two horse power; we have hitherto learnt nothing of the result, of this proposed experiment; and suppose, therefore, that the trial has not yet been made.

Since publishing the article above alluded to, it has appeared to us that should a much less power be attained by such a machine than that which is now sought for, say the power of man only, it would still be equally valuable with the steam engine, and would produce as great, if not a greater change, in the economy of the useful arts, as has been produced by that instrument; this, however, is under the proviso that the cost of materials consumed in performing the work of a day should be less than that given for the labour of a man. Who is there who would not, under such circumstances, need such a machine?

If we hire a man by the day we must not allow him to be idle, as in that case we give our money for nothing. The current of his life flows on, and he must be fed and clothed or the stream will stop. But give us a machine which is not costly at first, and if it works but one hour in the twenty-four, will itself be a consumer in that proportion only; a machine which we can at any moment set to turn our lathes, our grindstones, our washing machines, our churns, our circular saws, and a catalogue of other things which it would be no easy task to make out; such a machine would also perform a million of other operations by the conversion of a rotary into a reciprocating motion; and we again ask who is there among us who would not want one? Our farmers, our mechanics, and our housekeepers generally, must also be supplied. We could no more submit to live without it, after it has been once introduced, than we can now submit to travel at the slow rate of ten miles an hour, an event which we have learnt to think one of the miseries of human life.

With such a machine at our command we should soon wonder how we could have lived so long without it; and if taken from us it would leave a most awful chasm in the necessaries of life, of the existence of which our fathers never dreamed, and which happily we could not be called upon to witness so long as the store house of nature would enable us to obtain zinc and sulphuric acid at a cheap rate.

The steam engine cannot be used to advantage where it has not the labour of several horses to perform, as, whether large or small, it requires the constant attention of the engineer, or of the fireman, and is kept at work at an expense which is relatively increased as its power is diminished. One giving the power of a man only would be employed, at a cost which would be incalculably increased; of course it is not, and never will be used, under such circumstances.

Let it not be said that we are prophesying about what is to happen; not so by any means; but be it remembered that we are speaking of what is a possible contingency. We have no doubt respecting the practicability of obtaining the power of a man by the agency of electro magnetism; we believe that such a machine may be kept at work without any considerable tax upon the time of the person using it, and we further believe that the only thing which can prevent its coming into use is, the cost of the materials employed in operating it; the statements which we have heard upon this point are extremely contradictory, and upon the whole, are far from encouraging; the time, however, is not remote when this point will be determined.

Notes and References

1. L. Pearce Williams, 'Coulomb and the impossibility of electromagnetism', *Contemporary Physics*, 1962, 4: 113–23, 113ff.

2. Hans Oersted, *Experimenta circa effectum conflictus electrici in acum magneticam*, (Hafniae, 1820).

3. Auguste de la Rive, 'Notice sur quelques expériences électro-magnétiques', *Bibliothèque universelle, science et arts*, 1821, 16: 201–3; Dominique Arago, 'Extrait des séances de l'Académie Royale des Sciences', *Ann. Chim. Phys.*, 1820, 15: 80; Humphry Davy, 'On the magnetic phenomena produced by electricity', in a letter from Davy to W.H. Wollaston, *Phil. Trans. R. Soc. Lond.*, 1821, 111: 7–19; J.S. Schweigger, 'Zusaetze zu Oersted's elektro-magnetische Versuchen', *Schweigger's Journal*, 1821, 31: 1–6; L. Gilbert, 'Untersuchungen ueber die Einwirkung des geschlossenen galvanisch-elektrischen Kreises auf die Magnetnadel', *Annalen der Physik*, 1820, 66: 331–91.

4. B. Gee, 'André Marie Ampère (1775–1836)', *Physics Education*, 1970, 5: 359–69.

5. See L.P. Williams in *Faraday Rediscovered*, eds D. Gooding and F.A.J.L. James, (London, 1985), ch. 5, particularly 91–6.

6. Joseph Agassi, *Faraday as a Natural Philosopher*, (Chicago, 1971), ch. 3, particularly 43–78; L. Pearce Williams, *Michael Faraday: A Biography*, (New York and London, 1965), 151 ff.; William Berkson, *Fields of Force*, (London, 1974), ch. 1, particularly 39 ff.; and David Gooding in *Faraday Rediscovered*, op. cit. (note 5), 104–35, particularly 110–20.

7. *Diary: Being the various philosophical notes of experimental Investigation made by Michael Faraday DCL*, 7 vols, (London, 1932–6), I, 49.

8. Ibid., I, 50 ff.

9. M. Faraday, 'On some new electro-magnetical motions, and on the theory of magnetism', *Quart. J. Sci.* 1821, 12; 186. Newman is cited as maker in an addendum in Faraday's *Experimental Researches in Electricity*, 2, (London, 1844), 147.

10. For Barlow, see *DNB*: sv; *DSB*: sv.

11. P. Barlow, 'Method of correcting local variation of ships' compasses', *Trans. Soc. Arts*, 1821, 39: 76.

12. P. Barlow, *An Essay on Magnetic Attractions*, 2nd edn, (London, 1823), 233 [and cited again in 'Account of a series of electro-magnetic experiments', *Edinb. Phil. J.*, 1823, 8: 368–82, 370].

13. Ibid. Barlow did not publish any conception of the galvanic fluid until later. [*Edinb. Phil. J.*, 1825, 12: 105–14].

14. See P. Barlow, op. cit. (12), addendum.

15. Ibid., 279.

16. Pixii is acknowledged by Ampère in 'Description d'un appareil électro-dynamique', *Ann. Chim.*, 1824, 26: 390–410, 392n.

17. William Sturgeon's active years occupy seven columns of *DNB*. Background account here is derived from J.P. Joule, 'The life and writings of the late Mr. William Sturgeon', *Mem. Manchr Lit. Phil. Soc.*, 1857, 14: 53–83.

18. Quoted by Joule (17), apparently from an MS. of Sturgeon, of which this part also appeared in *Ann. Electr.*, 1842, 8: 81.

19. That Sturgeon gleaned ideas from Woolwich is clear from later acknowledgements such as: 'Mr Christie very politely showed me several of his experiments at the time he was carrying them on; and when published, presented me with a copy of the paper in which they were described. In a very short time repeated some of those experiments . . .'. Sturgeon, 'On the distribution and retention of magnetic polarity in magnetic bodies', *Phil. Mag.*, 2nd ser., 1832, 11: 270–9.

20. [Sturgeon], 'Papers in chemistry, no. III. Improved magnetic apparatus', *Trans. Soc. Arts*, 1825, 43: 37–51, 38 ff.

21. Ibid., 39.

22. Ibid., 41.

23. In addition to the better known establishments such as the Russell Institution or the City Philosophical Society there were some ten mechanics' institutes in the London region alone by 1828. T. Kelly, *George Birkbeck: Pioneer of Adult Education*, (Liverpool, 1954), appendix IV.

24. J. Cumming, 'On the application of magnetism as a measure of electricity', *Trans. Camb. Phil. Soc.* 1822, 1: 269, 281–6; M. Faraday, op. cit. (9), and *Quart. J. Sci.* 1822, 12: 283, 416–21; M. Tatum, 'On electro-magnetism', *Phil. Mag.*, 1823, 61: 241–3; 1823, 62: 107–9; *Mechanics Register*, 1825, 54: 41.

25. Peter Barlow's London Institution course is cited (but with no details) in P.M. Roget, *Electricity, Galvanism and Electro-Magnetism*, (London, 1832), 22n.

26. Little is known of Sturgeon's Addiscombe lectures other than occasional glimpses through his published *Lectures on Electricity*, (London, 1842).

27. 'A letter from M. Ampère to M. Gerhardi on various electro-dynamic phenomena', and 'Sequel of the memoir of M. Ampère on a new electro-dynamic experiment', *Phil. Mag.*, 1825, 66: 368–73, 373–87 and 1826, 67: 37–45; 'Electricity and magnetism', *Mech. Mag.*, 1827, 7: 52 f., 67–9, 95 f., 103 f., 128 f. and 133–5, respectively.

28. J. Newman, *Catalogue of optical, mathematical and philosophical instruments manufactured and sold by ——, 122 Regent Street*, (London, 1827), 20 ff.

29. The *Minutes* of the Committee of the London Mechanics Institute for 12 November 1827 record: 'that Mr Watkins of Charing Cross had present an instrument of his construction for producing the rotation of electrical wires by electromagnetism.' I am indebted to Dr R.E. Swainson, Secretary and Clerk to the Governors, Birkbeck College (London) for assisting in the location of this detail. Private correspondence: 28 June 1985.

30. F. Watkins, *A Popular Sketch of Electro-magnetism, or Electro-dynamics; with plates of the most approved apparatus for illustrating the principle phenomena of that science and outlines of the parent sciences, Electricity and Magnetism*, (London, 1828), 12.

31. *Description of apparatus constructed and sold by Watkins and Hill, for illustrating the most striking phenomena of electro-magnetism*, in Watkins's, *Popular Sketch*, (30), 70–83. Watkins and Hill, *A descriptive catalogue of optical, mathematical, philosophical,*

and chemical instruments constructed and sold by ——, Curator of Philosophical Apparatus in the University of London, 5 Charing Cross, (London, 1832), 61.

32. Watkins, *Popular Sketch*, (30), iv.
33. [Review], *Phil. Mag.*, 2nd ser., 1828, 4: 220–2, and [New Publications], *Mech. Mag.*, 1829, 11: 29–31, 45–8.
34. Ibid., 221 and 29, respectively.
35. Ibid., 221 and 48, respectively.
36. Further discussion of this critique of Watkins's text is given in B. Gee, *The place and contribution of the instrument maker in scientific development, 1820–1850, with special reference to electromagnetism and the diffusion of science*, (PhD thesis, Leicester University, 1988), 62–7.
37. Watkins's *Popular Sketch* was, however, reprinted in 1856 with updating footnotes by the brothers Elliott, who succeeded to the old Watkins and Hill establishment.
38. See D.L. Emblen, *Peter Mark Roget: The Word and the Man*, (London, 1970).
39. Watkins & Hill, *Catalogue*, (31), passim.
40. P. Barlow, 'A curious electro-magnetic experiment', *Phil. Mag.*, 1822, 59: 241 ff.
41. P.M. Roget, *Electricity, Galvanism, and Electro-Magnetism*, (London, 1832), 20.
42. Ibid., 23.
43. *Ann. Chim. Phys.*, 1821, 18: 331.
44. Marsh [communicated by P. Barlow], 'On a particular rotating cylinder', *Phil. Mag.*, 1822, 59: 433–5 + Plate V.
45. *Phil. Mag.*, 'Intelligence', 1823, 62: 237.
46. W. Sturgeon, 'Experimental and theoretical researches in electricity, magnetism, &c.', *Ann. Electr.*, 1842, 8: 81 ff.
47. P. Barlow, 'An account of some electro-magnetic combinations for exhibiting thermo-electric phenomena, invented by Mr. James Marsh of Woolwich; with experiments of the same by Peter Barlow', *Phil. Mag.*, 1823, 62: 321–27.
48. W. Sturgeon, 'Description of a rotative thermo-magnetical experiment', *Phil. Mag.*, 1824, 63: 269–71.
49. Ibid., 270.
50. Ibid.
51. F. Watkins, *Popular Sketch*, (30), 55.
52. *Mech. Mag.*, 1829, 11: 47.
53. W. Sturgeon, *Phil. Mag.*, 1832, 11: 202.
54. Gerrit Moll's electromagnetic interests are covered by H.A.M. Snelders in *Ann. Sci.*, 1984, 41: 35–55.
55. G. Moll, 'Electromagnetic experiment', *Am. J. Sci.*, 1830, 19: 329–37. This paper contains an interesting survey of some of the great magnets in earlier times.
56. Ibid., 333.
57. For Henry-Moll links see N. Reingold, *Science in Nineteenth Century America*, (New York, 1967), 62–76.
58. J. Henry, 'On the application of the principle of the galvanic multiplier to electro-magnetic apparatus, and also to the development of great magnetic power', *Am. J. Sci.*, 1831, 19: 400–8. These results are all the more remarkable when it is remembered that he was ignorant of Ohm's work on conductivity. In 1833, he asked Bache: 'Can you give me any information about the theory of Ohm? Where is it to be found?' [See Mary A. Henry in *Electr. Engineer*, 1893, 13: 28]. It was not until 1837, during his European tour, that he became acquainted with this theory.
59. J. Henry and ten Eynck, 'An account of a large electro-magnet made for

the laboratory at Yale College', *Am. J. Sci.*, 1831, 20: 201–3.

60. J. Henry, 'On a reciprocating motion produced by magnetic attraction and repulsion', *Am. J. Sci.*, 1831, 20: 342.

61. T. Edmondson Jr, 'The Rotating Armatures', *Am. J. Sci.*, 1834, 26: 205 ff.

62. N. Reingold, *The Papers of Joseph Henry*, vol. 3, (Washington, 1979), 447 n.

63. Ibid., 446 ff.

64. Silliman's editorial remarks in 'Notice of the electro-magnetic machine of Mr. Thomas Davenport of Brandon, near Rutland, Vermont', *Am. J. Sci.*, 1837, 32: 3 ff.; and the (anonymous) compilation under [Davenport, Thomas], *Electro-magnetism: History of Davenport's Invention of the Application of Electro-magnetism to Machinery . . .*, (G. & C. Carville & Co., Geo. E. Hopkins & Sons, 1837), 14 n.; for Davenport's own descriptions see *Electr. Engineer*, 1891, 11: 4; finally see M. Somerville, *Electromagnetism—History of Davenport's Invention of the Application of Electromagnetism in Machinery*, (New York, 1837).

65. N. Reingold, (62), 445 ff.

66. [Sturgeon], 'Historical sketch of the rise and progress of electro-magnetic engines for propelling machinery', *Ann. Electr.*, 1839, 3: 429–39; F. Watkins, 'On electromotive machines', *Phil. Mag.*, 1838, 12: 190. Robert Hare wrote to Henry saying: 'You will observe in the last number of the *Philosophical Magazine* Watkins does you justice as the original inventor of a machine for producing power by electromagnetism. In what he alleges I fully concur.' Reingold, vol. 4 (1981), 36. A letter from Henry to Watkins, dated 19 June 1839, contains the following remark: 'I am much indebted to you for the very flattering notice you gave in the Phil. Magazine of my little vibrating machine and hope to have frequent opportunities of returning the compliment relative to your various labours in the way of science.' From a copy of a letter retained by Henry and reprinted in N. Reingold, (62), 4 (1981), 240 ff.

67. It would be difficult to attribute an absolute order of priority although the following key names and dates indicate some of the other developments: Shulthouse in Zurich (1833); H. Jacobi at Paris and Dal Negro at Padua (1834); Callan at Maynooth, J.W. McGaulay of Dublin and Watkins in London (1835); C. Page of Salem, USA (1836–7); Davenport (1834–7); Sturgeon (1832–7); Watkins (1838); and J.P. Joule of Salford (1839). See [Sturgeon] (1839), (66), and Franklin L. Pope, 'The invention of the electric motor', *Electr. Engineer*, 1891, 11: 1–5.

68. W. Ritchie, 'Experimental researches in electro-magnetism and magneto-electricity', *Phil. Trans. R. Soc., Lond.*, 1833, 123 (Part II): 313–21.

69. D. Davies, *Manual of Magnetism*, (edition 1857), 212.

70. W. Sturgeon, 'Description of an electro-magnetic engine for turning machinery', *Ann. Electr.*, 1837, 1: 75–78.

71. Ibid.

72. W. Sturgeon, (1839), (67), 436 ff.

73. *A descriptive catalogue of optical, mathematical, philosophical, and chemical instruments and apparatus sold by Watkins and Hill, 5 Charing Cross*, (London, 1838), 86. This catalogue, like the previous 1836 edition, retained the Saxton magneto-motive machine (£4.4*s*.) and the cheap Ritchie machine (15*s*. to £1. 1*s*.).

74. E. Palmer, *New Catalogue, with three hundred engravings of apparatus . . . manufactured and sold by him at 103 Newgate Street*, (London, 1840), 30.

75. F. Watkins, 'On the magnetic powers of soft iron', *Phil. Trans.*, 1833, 333–42.

76. F. Watkins, 'On magneto-electric induction', *Phil. Mag.*, 1835, 7: 107–13.

77. F. Watkins, 'On electro-magnetic motive machines', *Phil. Mag.*, 3rd ser., 1838, 12: 190–6, + plate 4.

78. No extension to Joule's well-documented researches is added here other than the note that his first electromotive engine could raise 15 lb. by 1 foot per minute using a battery of forty-eight 4 square inch zinc plates. [*Ann. Electr.*, 1838, 2: 122 and 1839, 3: 437 ff.].

79. F. Watkins (1838), (77), 193.

80. Ibid., 194.

81. Ibid.

82. Ibid., 195.

83. Ibid.

84. For example, see Anthony Hyman, *Charles Babbage*, (Oxford, 1982), passim.

85. F. Watkins, (1838), (78), 196.

86. M.H. von Jacobi, 'On the application of electro-magnetism to the moving of machines' from *L'Institut*, 1834, 2: 394 ff. and translated in Sturgeon's *Ann. Electr.*, 1837, 1: 408–15, 419–44. This source would have been the source of Watkins's information and caution. Note: Moritz Hermann von Jacobi, brother of the famous mathematician Karl Jacobi, was living in Königsberg and practising as an architect when he presented his mathematical analysis to the Paris Academy in May 1834. Interestingly, this date precedes Davenport's July 1834 demonstration and therefore he may hold priority over him.

87. M. Jacobi, 'Ueber die Principien der elekromagnetischen Maschinen', *Annalen der Physik*, 1840, 51: 358–72. See also D.V. Efremov, *The Electromotor and its Historical Development*, (Moscow and Leningrad, 1936), 230–47 (in Russian).

88. Reingold, vol. 4, (1981), (62), 268.

89. Henry Noad [in *Lectures on Electricity*, (2nd edition, London, 1844), 359] talks of the design of such engines as a problematic desideratum in 1844. For G. Knight Jr, see *Catalogue of the Illustration and Manufactures, Inventions and Models, Philosophical Instruments, . . . of the second exhibition of the BAAS*, August 1839.

90. J.C. Michalowicz, 'Origin of the electric motor', *Electr. Engineering*, 1948, 67: 1035–40.

91. *Exhibition of the Works of Industry of all Nations 1851, Report by the Juries* (London, 1852), 282. For details of Soren Hjorth's first machine, see *Electrician*, 1882, 9: 173–5.

92. Thomas P. Jones was Editor at this time but this piece is likely to have been written by his assistant, John Griscom.

Teaching Telegraphy and Electrotechnics in the Physics Laboratory

William Ayrton and the Creation of an Academic Space for Electrical Engineering in Britain 1873–1884

GRAEME GOODAY

When the electrical engineer feels himself full of pride at the greatness, the importance, and the power of his industry, and when he is inclined to think slightingly of the deflection of a little magnet compared with the whirl of his 1,000 horse-power dynamo, let him go and visit a certain dark store room near the entrance hall of the Royal Institution, and, while he looks at some little coils there, ponder the blaze of light that has been spread over the entire world from the dimly lighted cupboard in which those dusty coils now lie. (William E. Ayrton, Presidential Address to Section A of the British Association for the Advancement of Science, 1898.)[1]

Electrical engineering is in a curious position. It owes its being altogether to scientific men, to the laboratory and desk-work of a long line of experimenters and philosophers. Even now the work going on in a laboratory today becomes the much larger work of the engineer tomorrow. When at length the laboratory experiment is utilized in engineering, we see that there is no other kind of engineering which so lends itself to mathematical treatment and exact measurement. (John Perry, Address to the Institution of Electrical Engineers, 1900.)[2]

Three months in the laboratory and a month in charge of an electric light machine, and then forth stepped the full-blown electrical engineer blushing with the consciousness of his new dignity. . . . Much as the various schools and technical colleges have done for the progress of the electrical industries, they

> would do still more if they could make every one of their pupils realize that a college course can never supplant the apprenticeship of the [work] shops, and can give but half a training—namely, that half which in turn the apprenticeship of the shops will not and cannot afford. ('Training for the electrical industry', editorial in *The Electrician*, 1890.)[3]

In the last two decades of the nineteenth century academic electrical engineers such as William Ayrton paid ritual homage to Michael Faraday's laboratory researches on electromagnetic induction as a major historical source for their new technologies of power machinery and lighting. Yet for the new generation of electrical engineers that appeared in Britain from 1878 onwards the privileged venue of electrical practice was the factory workshop, power generating plant or public installation site, and *not* the institutional laboratory. Moreover, since the mid-1850s the pre-existent community of telegraph engineers, who constructed and operated the global network of cables, had daily utilized knowledge of the electrostatic retardation of submarine signals, which Bruce Hunt has shown to have originated not in Faraday's laboratory but in Clark and Statham's manufacturing test-room.[4] How then did William Ayrton establish the academic laboratory as a legitimate space for training electrical engineers outside the more familiar domains of professional practice?

This paper will document how Ayrton negotiated the 'half' of the electrical engineer's training that the workshops would not and could not 'afford'. It will show how, with the support of colleagues such as John Perry and trade journals such as *The Electrician*, he created an efficient laboratory regime of professional training at Finsbury Technical College by 1884.[5] Further, it will illustrate how Ayrton, who began his career as a practising telegraph engineer, won public assent for this scheme by maintaining a fastidious deference to the orthodox rites of apprenticeship under the master engineer: this counted as the other 'half' of a novice telegraphist's training, following the powerful traditions of British civil and mechanical engineering.[6] To complement the special workshop routines in which telegraphists learnt the 'craft' of making cables and equipment in commercial workshops and the heroic endeavours by which they learnt the 'arts' of laying, operating, testing and repairing cables in the diverse environments of the British Empire and beyond,[7] the regime which Ayrton created was one of laboratory precision measurement[8] that drew heavily upon the contemporary culture of academic experimental physics.[9]

Valuable historical scholarship already exists on some aspects of Ayrton's work in relation to the evolution of electrical engineering and its training methods. D.W. Jordan has discussed the role of the City and Guilds in sponsoring 'technical education' through their establishment and maintenance of the Finsbury Technical College from 1879, the Central Institution at South Kensington from 1885, and the contemporaneous administration of their distinctive Technological Examinations.[10] W.H.

Brock has discussed the Tokyo context of Ayrton and Perry's collaboration in the practical teaching of academic engineering to Japanese students during the 1870s, and also the construction of the laboratories that were built for them and their chemist colleague H.E. Armstrong at Finsbury Technical College between 1881 and 1884.[11] Rollo Appleyard's classic history of the Institution of Electrical Engineers (IEE) discusses the legislation of the IEE on the proper training of an electrical engineer from its foundation as the Society of Telegraph Engineers in 1871 until 1931; it has recently been supplemented by the celebratory account of the IEE by Reader *et al.*[12]

None of these accounts, however, discusses the *problematic* status of the laboratory as an innovation in the training of telegraph engineers and the later generations of electrical engineers. In discussing Ayrton's work I shall therefore establish first the explicit hostility shown in the early 1870s towards the academic domestication of telegraphy by two Scottish University engineers directly or indirectly involved with Ayrton's career: Fleeming Jenkin and W.J.M. Rankine. Second, I shall discuss the parallel but permeable social demarcation lines between academic physics and practical telegraphy with reference to Ayrton's intermittent experience of London during the 1870s: the small overlap in metropolitan membership between the Physical Society and the Society of Telegraph Engineers included all the major physicists who taught telegraphic skills in their laboratories, such as Ayrton from 1877. Third, I shall discuss Ayrton's creation of a telegraphic laboratory during his sojourn in Japan from 1873 to 1878 and contrast its effective internal functioning with the problems he encountered in translating his student's expertise into Japanese culture. I shall conclude with a discussion of how Ayrton created a London laboratory for training telegraphists and electricians in the early 1880s, how he structured its operation to avoid disruption of the immediate cultural milieu of British industry, and how he overcame very public problems in establishing that this laboratory was an effective vehicle for teaching the professional skills of electrical practice.

As a sub-theme in this analysis I shall allude to the 'irony' of Michael Faraday—laboratory experimenter *par excellence* of the nineteenth century—being invoked as the 'founding father' of electrical engineering. His status as such is evident from the iconography employed by the Institution of Electrical Engineers (see frontispiece). It is explicit in S.P. Thompson's[13] hagiographical preface to *Michael Faraday, His Life and Work* in 1898, in which the author granted his subject the Whig accolade of a lifetime spent in 'laying the foundations . . . for great developments of electrical engineering of the last twenty years.'[14] Davy, Oersted, Wheatstone, Cook, Ronalds, Thomson, Maxwell, Edison, Bell and many others might have been selected to play such a paternal role; yet soon after his death in 1867, this specific institutionalization of Faraday was manifest, for example, in the BAAS unit of electrical capacitance, the 'farad', and in the christening of William Siemens's cable-laying ship as the *SS Faraday*.[15] The persistent invocation of Faraday in electrical trade journals throughout the 1870s and 1880s,[16] illustrated tangentially in the course of this paper, will be taken as an index of how far practitioners of electrical engineering found it expedient to use the moral pedigree of laboratory physics to legitimate their new profession. Given this usage of Faraday's reputation, it will become clear during the course of this paper why the context in which electrical engineering was first manifested as an acedemic subject should have been Ayrton's City and Guilds of London laboratory of 'technical physics'.

THE PLACE TO LEARN PRACTICE: THE TERRITORY OF 'ART' IN TELEGRAPHY

> I have the strongest feeling against any attempt to substitute collegiate training for practical apprenticeship. So far as colleges attempt to teach practice they are a sham. . . . What has been taught by practice must still be taught by practice. The business of the school is to teach those things which practice in an art will not teach a man.[17]

Much scholarly attention has been directed at the way in which the laboratory practices and instrumentation of William Thomson were imported into the (successful) laying of the 1866 Atlantic Telegraph Cable.[18] Historians of technology have said less about how incongruous Thomson's earlier participation in the abortive cable-laying expedition of 1858 appeared to contemporary electricians, who had vastly greater experience of laying and operating cables. William Ayrton, as one of Thomson's former students, once remarked that Thomson's intransigent demands in 1858 for rigorous laboratory resistance measurements to monitor cable quality and his insistence upon the use of his delicate mirror galvanometer for receiving telegraph signals 'appeared to the electrician [E.O.W. Whitehouse] as arising from the ignorance of an inexperienced young man who had never erected a mile of telegraph line in his life, and would not have been given a job in any telegraph office.'[19]

Yet, however much ignominy was placed upon Whitehouse's conventional 'rule-of-thumb' methods for the failure of the 1858 cable, few have commented upon the enduring strength of the 'craft' tradition in British telegraphy that survived even this controversial episode. Specifically, no historians have documented the British telegraphists' distinction between the laboratory-generated 'science' of their profession and the practical 'art' of applying their expertise to the natural world beyond the laboratory walls. As the editor of the *Telegraphic Journal* put it in 1876:

> Faraday determined the *science* of magneto-electricity when he discovered the fact that currents can be produced by the motion of a wire in a magnetic field. . . . Ohm laid the foundation of the *science* of electric currents when he developed the laws of resistance, and [Cromwell] Varley developed the *art* of testing when he applied these laws to determine the distance from the shore of the broken end of a cable.[20]

Irrespective of the institutional contexts in which Ohm and Faraday established these usages, by the conventions of British telegraphy novice practitioners were required to learn the 'arts' of Varley and his telegraphic colleagues in an 'apprenticeship' or 'pupilage' under the master engineer in the 'workshop' and the 'field'. From the late 1860s the wealthier prospective telegraphist from a middle-class milieu might have been able to learn the 'science' of telegraphic measurement in the lecture theatres and physics laboratories of London and Glasgow in order to pass Civil Service examinations and thenceforth enter the Government Telegraph Service.[21] Although his social position might also have enabled him to bypass the full apprenticeship process, he was also required, like his poorer contemporaries, to spend at least some time learning the 'arts' of manufacturing, laying, operating and testing telegraph cables under an established practitioner.

William Ayrton's early life as a telegraph engineer must be seen in the context of this conventional territorial separation between the 'science' and 'art' of his profession, for, although highly singular in its later course, it exemplifies some of the typical training and career trajectories of a young practitioner in the 1860s and 1870s. In 1864 Ayrton began his scientific education at University College London. In his final year, 1866–7, he studied in the newly opened physics laboratory while also acting as assistant to Professor George Carey Foster. Simultaneously graduating as a BA (Hons.) from the College and coming first in the examinations for the Indian Government Telegraph Service, Ayrton was despatched by the Secretary of State for India to study in Sir William Thomson's laboratory at the University of Glasgow during the winter session of 1867–8.

After this Ayrton was sent for practical experience at the works of the Telegraph Construction and Maintenance Company and for training in the Post Office protocol of telegraphy under William Preece at Southampton. In September 1868 he travelled to Bombay to take up his post as a fourth-grade assistant superintendent. Under the aegis of the electrician C.L. Schwendler, Ayrton and his subordinates transformed the Imperial operation of Indian telegraphy by instituting a system (derived from Varley's submarine methods) of determining the position of, and subsequently rectifying, faults in the cable network of the subcontinent.[22]

After being made electrical superintendent and taking charge of telegraph offices in Bombay and Calcutta, Ayrton returned to England in 1872 and worked both for the India Office in testing cable insulation, and for William Thomson and Fleeming Jenkin in supervising electrical testing for the Great Western Telegraph Company. Ayrton's next career move was to take up the Professorship of Natural Philosophy and Telegraphy at the Imperial Engineering College in Japan in 1873, where he created his prototype of a telegraphic laboratory.[23]

It is important to analyse the contextual objections to such an innovation both from Jenkin and from Thomson's close colleague W.J.M. Rankine in terms of their traditionalist views of the appropriate balance between their academic teaching of engineering theory and apprenticeship training in the 'art' of telegraphy, since Ayrton would have encountered these through his employment and by studying at Glasgow in 1867–8.[24] Jenkin and Rankine held professorial teaching appointments in engineering at the Universities of Edinburgh (1868–85) and Glasgow (1855–72) respectively and were concurrently active in engineering practice—Jenkin with the Atlantic Telegraph Company and Rankine as consulting engineer for the Glasgow shipping industry as a close associate of Lewis Gordon, his Glasgow predecessor and then chief electrician of the cable manufacturers R.S. Newall and Co.[25] Both had been apprenticed in a conventional manner earlier in their careers as civil engineers, and so although they supported the college training of the aspiring engineer in the scientific theory of his profession, they none the less opposed any attempt to train students in any part of the practice of engineering, particularly, in Jenkin's case, telegraphic engineering. We can see this from the manner in which Jenkin and Rankine defended the integrity of the apprenticeship against the stratagems of laboratory propagandists to align telegraphy engineering with their expanding empire of academic culture. This essentially territorial conflict can be seen most explicitly when Jenkin and Rankine were interviewed by the Royal Commission on Scientific Instruction and the Advancement of Science in the early 1870s.

From 1870 to 1875 the Royal Commission, headed by the Duke of Devonshire, was engaged in a massive survey of British scientific education and research. As a member of this Commission and an ardent propagandist for laboratory teaching,[26] T.H. Huxley attempted to coax his interviewees into expressing agreement with his pro-continental views on technical education. For example, he attempted to force Jenkin and Rankine to assent to his view of an exclusively college-based training for engineers on the French and German model as a 'superior' alternative to apprenticeship. To Huxley's suggestion that not all such technical education was 'bad and useless' Jenkin gave very short shrift indeed, replying that 'if you mean by technical education, attempting to teach a man his business by a college course, I think it is a very mischievous delusion indeed.'[27] Huxley's next tactic was to assert that a college engineering course would have as much *practical propriety* as the hospital training of a physician; Jenkin, however, denied such a professional parity:

> in a college . . . a medical man . . . does find some real practical work, not perhaps to do, but to see very closely in the hospital. [Now] if I could concentrate railways, canals, harbours, roads and telegraphs into a kind of engineering hospital, I would not object to a technical engineering college, but I do not think you can have that.[28]

Questioning more specifically, Huxley asked, 'Would your objection apply equally to telegraphy. It would be possible to have a telegraphic hospital, would it not?' To this Jenkin replied:

> I think not. You may have a piece of wire really 600 miles in length [a typical length of international telegraph cable] represented, as regards certain phenomena, in a little box, but you cannot represent

> the thing itself. A museum of telegraph posts and wires and bits of cable are very poor representatives of the *real* thing.[29]

Jenkin differentiated the two cases by explaining that in a hospital 'the medical man sees the actual operation at any rate, but he does see the true thing, whereas the engineer at a college does not see the true thing. Even where attempts are made to approximate to the workshop, as in the École des Arts et Métiers [Paris], it is not the true thing.'[30] Attempting to outmanoeuvre Jenkin in his denial of the cognitive parity of hospital and college, Huxley then goaded Jenkin to explain why an attempt at engineering work could not be 'true' or 'real' in the college environment.

Jenkin replied by explaining the traditional manner in which the pupil or apprentice learnt his trade practices by *emulating* the master engineer in the context of the engineering works; he gave a standard example of testing cable quality by measuring its resistance:

> I and my principal assistant test the cable together, and my pupil looks on, he perhaps puts down some figures, or asks for a little explanation. The next time that pupil has to test a cable he does not [yet] know all about it, but [at least] he has seen the real thing in a totally different way from that which can be produced in the lecture room of a college.[31]

Jenkin's sole criterion of 'authentic' training in telegraphic measurement practices was that it was learned in the professional context of practice by emulation of the practising engineer—college work did not have the necessary 'contextual integrity' in this respect.[32]

In his testimony to the Royal Commission two years later, W.J.M. Rankine was also explicit about the futility of an imitation engineering workshop in the science college. Unlike Jenkin, however, he differentiated the 'integrity' of the engineering site works as a matter of technical *scale*:

> I think it is useless to try and give instruction in practice proper in a University. If for instance we set up a mechanical workshop with machine tools in it in our university, or if we got, say, a mile or two of railway, and set our students to superintend the works, this would be worse than useless. The difference between doing things on a small scale like that, and doing things on a great scale, as in actual practice, is so great that the students would only be led to fancy that they had more knowledge than they really did possess, and therefore, I am for having the practical knowledge acquired at separate times and places by practice on a great scale.[33]

Rankine argued that the instruction in 'mechanical manipulation' available in the laboratory of his Glasgow University colleague, William Thomson, was 'useful' to prospective engineers but *only* as a training in building instruments for experimental physics. As Rankine put it,

> This would be useful to young engineers as well as to other students; but they should not be led to suppose that in practising mechanical manipulation that they are at all qualifying themselves for the practice

on the great scale that one meets with in engineering workshops, or in the execution of lines of railway or other engineering works.[34]

We can see that within the Scottish engineering traditions with which Ayrton had close contact in the late 1860s and early 1870s there were two crucial features of authentic engineering training that could not be replicated in academically domesticated training in engineering practice: the *contextual emulation* of the master engineer, and the *integrity of scale* of surrogate practices.

Returning to Jenkin, we can surmise the relationship between his engineering teaching and the kind of measurement-oriented courses taught to aspiring telegraphists by William Thomson at Glasgow, G.C. Foster at University College London and William Grylls Adams of King's College London.[35] Bernard Samuelson asked Jenkin, 'Are you aware that there is a great confusion in men's minds between such special courses as yours, and laboratory instruction in colleges?' Jenkin replied, 'I think that there is a good deal of confusion. I think that people, however, are coming to understand it better.'[36] It is important to emphasize that the functional divergence between the academic physics laboratory and the industrial workshop was not local to the Scottish context. In fact we can see that in the 1870s this was institutionalized in the differences of interest and practice between the London-based Society of Telegraph Engineers and the Physical Society; there was, however, a very small—but for our purposes very significant—common membership of these two societies.

THE PHYSICAL SOCIETY AND THE SOCIETY OF TELEGRAPH ENGINEERS

Within six years of Faraday's death two societies directly concerned with 'electrical science' were founded in London. The Society of Telegraph Engineers (STE), founded in 1871, and the Physical Society, launched in 1873, institutionalized the distinctive but permeable domains of telegraphy and academic physics that coexisted in the capital. Reader cites Siemens's inaugural address as the first President of the STE, pointing out 'the remarkable fact that the manufacture of insulated wire and of submarine cables, is almost entirely confined to the banks of the Thames', and elsewhere I have shown that, contemporaneously, London had a higher density of academic physics laboratories than the centre of any other city in 1873: University College, King's College and the Royal School of Mines.[37] While the activities of the STE have been documented by both Appleyard (1939) and Reader *et al.* (1987), no equivalent account exists for the Physical Society.[38] Furthermore, Reader omits the Physical Society from his analysis of the way in which membership of the STE overlapped with London-based scientific societies. He follows Appleyard in interpreting the Physical Society as an association of 'professorial' scientists that operated behind a 'courteous wall' that demarcated its operation from the STE's institutional representation of telegraph engineers.[39]

Nevertheless, the comparative analysis of the STE and Physical Society presented below shows that between the 1870s and early 1890s such a clear demarcation did not in fact exist. Certainly contemporary commentators viewed them as being closely related in their subject areas: the proceedings of both the Physical Society and STE were, for example, conventionally reported side by side and with equal emphasis by trade journals such as *The Electrician*, the *Telegraphic Journal and Electrical Review*, *The Engineer* and *Engineering*.[40] Specifically I shall show that a small number of academic physicists interested in telegraphy played major roles in both and that these individuals were those primarily involved in the laboratory education of physics students and telegraphic/electrical engineers: George Carey Foster, William Grylls Adams, Sir William Thomson and William Aryton.

Since Appleyard and Reader give comprehensive documentation on the formation of the Society of Telegraph Engineers it will suffice here to relate their accounts to the discussion of traditions of telegraphic education above. The Society was created in 1870–1 following a six-year period in which a global telegraphic network was established across the Atlantic, India, Europe and Asia, largely constructed with British-controlled materials and expertise. Reflecting the strong Imperial interest in telegraphy, the STE was launched by a coterie of five military telegraph engineers in collaboration with two practising civilians and the retired practitioner E.O. 'Wildman' Whitehouse, who acted as chairman of the preliminary meetings.[41] Although Whitehouse's reputation as an electrician was virtually demolished as a result of his disastrous management of the 1858 Atlantic Telegraph Cable,[42] he practised for the last time as consulting electrician during the laying of the Malta–Alexandria cable in 1861 and then apparently devoted 'considerable attention' to such devices as 'electrical checks in connection with passenger fares in tramways and omnibuses.' Yet since Whitehouse was also apparently 'the cause of many others well known in submarine telegraphy adopting that profession', his functions in the nascent society were probably diplomatic and social in character and it is likely that his former protégés were a ready source of membership for it.[43]

While the participation of Whitehouse is a clear index of the traditionalistic alliances cultivated in the formation of the society, more dominant in the subsequent running of the STE was the paternalistic role of its senior ally, the Institution of Civil Engineers (ICE). By furnishing the STE with the free use of its chambers and meeting rooms the well-established ICE put itself in an optimum position to influence the activities of the new society.[44] Buchanan has characterized the ICE in this period as staunchly conservative in its policies on the proper training of the practising engineer, recommending apprenticeship or pupilage rather than scientific education, and wielding its authority to discourage deviant views among younger organizations such as the STE.[45] Appleyard emphasizes the extent to which the regulations of the STE were modelled upon those of the parent institution,[46] a prospective member being required to meet one of the following three conditions:

(a) He shall have been regularly educated as a Telegraph Engineer, according to the usual routine of pupilage, and have had subsequent employment for at least five years in responsible situations.

(b) Or he shall have practised on his own account in the profession of a telegraph engineer for at least two years, and have acquired a degree of eminence in the same.

(c) Or he shall be so intimately associated with the science of Electricity or the progress of Telegraphy that the Council would consider his admission to Membership would conduce to the interests of the Society.[47]

The STE thus legislated on the rights of different social groups to participate in the practice and discussion of telegraphy so as to marginalize academics and to prefer those active in the domain of commercial practice. Statistics given by Reader illustrate the dominance of practising telegraph engineers and manufacturers—as defined in categories (a) and (b)—in the first decade of the society: 87% in 1872 and 88% in 1881.[48] Eminent practitioners, such as Latimer Clark, William Preece and Cromwell Varley, and major industrialists, such as the Siemens brothers, were among those originally invited to join the society in January 1871.[49]

By contrast no academics were initially invited to join the Society under regulation (c), and only one of those elected to the first membership list in May 1871 was a Professor of Physics: G.C. Foster of University College, London.[50] By the end of 1872, however, the Council had evidently considered it expedient to admit reputable scientific figures, for the London contingent was by then expanded by the admission of Prof. W.G. Adams from King's College and Prof. John Tyndall of the Royal Institution, London; Prof. W. Thomson of Glasgow University was also admitted, along with his laboratory assistant J.T. Bottomley to join the Edinburgh-based Fleeming Jenkin as the axis of the small Scottish group. During 1874 the Cavendish Professor of Experimental Physics at Cambridge, James Clerk Maxwell, was similarly elected.[51] Reader's figures show, however, that the academic faction never constituted a large faction of the Society's membership: 5% in 1872 and 2% in 1882.[52]

These practitioners of academic laboratory physics in the STE closely identified their interest in techniques of accurate measurement with one specific concern of telegraph engineers: the technology of precise electrical testing.[53] One pertinent example of this concern can be seen in William Ayrton's first paper to the society in 1873. Writing as a young telegraph engineer recently returned from Imperial service in India, he related the techniques he had helped to develop and implement in measuring the resistance of sections of cables covering the subcontinent. Ayrton detailed the comprehensive regime of precision testing used to detect and locate faults in cables with an expedition commensurate with the requirements of efficient colonial communication and control. The subsequent discussion of his paper largely consisted of his former instructor W.H. Preece establishing that while the same testing devices were employed in both England and India, i.e. Wheatstone bridges and Thomson mirror galvanometers, the contrasting environmental conditions of climate, geology and culture entailed very different modes of practice.[54] Foster and Adams did not generally contribute to such characteristic discussions of the context of telegraphic practice at Society meetings, presenting instead papers on laboratory measurement technology: in 1872 Foster introduced to the Society what became his eponymous adaptation of the Wheatstone (resistance measurement) bridge, and in 1873 Adams presented a detailed commentary on Latimer Clark's methods of measuring potential difference.[55]

At about the time that the latter paper was being delivered Foster and Adams were assisting their close ally and colleague in the London triangle of academic physics laboratories, Frederick Guthrie of the Royal School of Mines, in the formation of a new organization: the Physical Society. Appleyard comments that news of this organization's first volume of *Proceedings* in 1874 led the Honorary Secretary of the STE, Frank Bolton, to move that a merger be effected to pre-empt the loss of 'its best members'. The STE Council formed a subcommittee with Foster among its number to investigate the prospects of such a merger.[56] Such political machinations evidently went no further because, presumably through Foster's mediation, it was soon apparent that the dominant constituency of the Physical Society was to be quite different: University, college and school teachers of physics, as well as non-aligned amateurs.

A Physical Society was first mooted by Frederick Guthrie, an academic laboratory physicist, in the late summer of 1873 following the unsatisfactory outcome of a recent dispute with the Royal Society over his electrical researches.[57] Since Guthrie's appointment as Tyndall's successor at the Royal School of Mines in 1869, his unorthodox researches had been the subject of some argument with the Secretary of the Royal Society, G.G. Stokes, particularly with regard to his idiosyncrasies of terminology and protocol.[58] Stokes generally refused to publish Guthrie's papers in the *Transactions*, relegating them to the *Proceedings*, much to Guthrie's chagrin.[59] Guthrie's final estrangement from the Royal Society occurred in 1873 when the same fate met his paper 'A new relation between heat and electricity'. Guthrie claimed to have discovered that white-hot metals lost negative charge more readily than positive charge, but referees of the paper differed. Fleeming Jenkin argued that the author had shown nothing 'more novel . . . than phenomena as well known as the discharge of a conductor and a point held opposite it'[60] and James Clerk Maxwell concluded that Guthrie's 'interesting experiments' related instead to the 'effects of heat upon air as altering its electrical properties', commenting that Guthrie should pursue these experiments further, 'more fully to work out the subject'.[61]

At the Jubilee Celebrations of the Physical Society fifty years later, it was not without irony that John Ambrose Fleming, an electrical engineer who had been a student of Guthrie's in 1872–3, remarked that 'Guthrie's observations were not capable of interpretation at the time he made them'. Retrospectively identifying a close connection between these controversial observations, Edison's work on carbon filament lamps and Fleming's own researches on 'glow lamps', Fleming argued that the 'effect of the loss of negative electricity from incandescent carbon in vacuo . . . was not . . . explained until the researches of Sir J.J. Thomson had made us acquainted with the electron and electron emission from incandescent bodies.'[62]

Whatever the conclusions reached by posterity on this episode, Guthrie's dismay at yet again having his work published only in the Royal Society's *Proceedings*[63] led him to abandon the Royal Society as a vehicle for his innovative work and instead canvass in 1873 for the formation of a new society. As Fleming put it, Guthrie

> thought that outside of complete or carefully worked out investigations in physical problems which might hope to claim a place in the *Proceedings* or the *Transactions* of the Royal Society . . . there might be a field for a Society which should encourage and publish accounts of physical researches of a less ambitious kind, and not refuse to accept descriptions or exhibitions of new experiments even if imperfectly explained or understood, provided they had interest and novelty for physicists.[64]

In the late summer of 1873 Guthrie issued a circular advertising this new society to all the major academic and amateur physicists in England,[65] and although Maxwell refused to join,[66] *Nature* reported of its first meeting on 29 November 1873 that 'most of the physicists of London were present'.[67] S.P. Thompson, an electrical engineer and former student of Guthrie's in 1873, pointed out in his address as President of the Society in 1901 that it 'was originated by teachers of physics [and] from its inception the Society has been actively supported by teachers of Physics in the Schools and Colleges of London, as well as by the Professors of Physics in the Universities and University Colleges of the United Kingdom, and by the Lecturers in Physics of the great Public Schools.'[68]

It was the interests of these people that were met by what G.C. Foster described as 'a society that took cognizance of smaller matters, points of technical detail, useful laboratory contrivances, experimental methods of illustrating physical principles, questions connected with methods of teaching.'[69] Accordingly the Society's meetings were uniquely characterized by experimental *demonstrations* that accompanied the reading of papers, Guthrie arranging for meetings to be held in his own laboratory complex at the Royal Schools of Mines; from 1874 to 1884 he acted as the experimental demonstrator for the Society.[70] The first paper accorded such treatment at the first full meeting of the society was an electrical paper by Guthrie's ex-student, J.A. Fleming, 'On the contact theory of the cell'.[71]

Russell Moseley has written disparagingly of the membership of the Physical Society, dismissing it on the grounds of Maxwell's refusal to join as merely an organization of '"second-order" experimental physicists' engaging in 'elementary scientific activities' that were 'unlikely to hold attention for the more senior academic physicist'.[72] This interpretation breaks down upon close scrutiny of the Society's membership and particularly its leadership: Dr John Hall Gladstone, a member of the STE and (shortly to be) Fullerian Professor of Chemistry at the Royal Institution, was strategically elected as President of the new Society in 1874. Gladstone, as symbolic heir of Faraday's institutional heritage, was succeeded in the Presidency by the most eminent of the laboratory-based physics professoriate of Britain, as can be seen from the following list:

1874-76	John Hall Gladstone	Royal Institution, London
1876-78	George Carey Foster	University College, London
1878-80	William Grylls Adams	King's College, London
1880-82	William Thomson	University of Glasgow
1882-84	Robert Clifton	University of Oxford
1884-86	Frederick Guthrie	Royal School of Mines, London
1886-88	Balfour Stewart	Owens College, Manchester.
1888-90	Arnold Reinold	Royal Naval College, London
1890-92	William Ayrton	Central Technical Institution, London[73]

Importantly, several of these men were also elected President of the Society of Telegraph Engineers (and Electricians)/Institution of Electrical Engineers: Thomson in 1874 and 1889, Foster in 1881, Adams in 1884 and Ayrton in 1892.[74] Ayrton was one of a handful of telegraphists who joined both societies in the 1870s, Cromwell Varley and Latimer Clark having been original members. Ayrton, however, departed for Japan before the Physical Society's first meeting in November 1873,[75] joining instead in 1877 during his last year as Professor of Natural Philosophy and Telegraphy at the Imperial College of Engineering in Tokyo—he had been Honorary Secretary for Japan of the STE since his arrival there. To comprehend Ayrton's entry into the constituency of this society it is necessary to see how, despite Jenkin and Rankine, he imported his expertise in outdoor telegraphy, along with his earlier experience of Thomson's and Foster's experimental instruction, into his management of the Tokyo teaching laboratories.

TELEGRAPHIC LABORATORIES IN AN ALIEN CULTURE: AYRTON IN JAPAN 1873-1878

> On looking over the collection of apparatus, much of which, however was in use all over the place, we felt that neither the Cavendish Laboratory nor the Oxford one, nor any others which it had been our privilege to see, could produce such experimental work as might come from this laboratory in Yedo, if only the men to use them were the same.[76]

> No wonder that [James Clerk] Maxwell jestingly said that the electrical centre of gravity had shifted to Japan.[77]

W.H. Brock has discussed in some detail the contextual connections in engineering education between the University of Glasgow and the Imperial College of Engineering established by the Japanese government in Tokei (Tokyo) in the 1870s. Numbered among the staff of the College, which was opened in makeshift accommodation during August 1873, were three engineers who, in addition to undergoing conventional apprenticeship or works experience, had close connections with the teaching of W.J.M. Rankine or William Thomson at the University of Glasgow. Henry Dyer, principal from 1873 to 1882, devised the six-year engineering curriculum that synthesized elements of both British and Continental training schemes, and also taught civil engineering. William Ayrton was Professor of Natural

Philosophy and Telegraphy from 1873 to 1878 and he was joined by John Perry as a second professor of engineering from 1875 to 1879 to form a long-lasting partnership in teaching and research.[78]

In authentic neo-colonialist fashion Dyer set out to 'experiment' with the Japanese students to create his ideal scheme of 'technical education'. Dyer, as Brock emphasizes, was one of W.J.M. Rankine's most successful engineering students at the University of Glasgow in the early 1870s, having previously been an apprentice for five years under the Glasgow marine engineer Alexander Kirk from 1863 to 1868.[79] What Brock does not discuss, however, is the way in which the scheme of technical education formulated by the 24-year-old Dyer consisted of scientific lectures and laboratory training during the months of the Scottish academic session, October to March, *and*, in authentically Rankinean manner, a *complementary* summer pupilage under a qualified practising engineer. The 1873 college programme laid out schemes common to civil, mechanical and telegraph engineering; architecture, applied chemistry, mining and metallurgy (Table 1).

Table 1 Dyer's 1873 Japanese college programme

	Year	*October to March*	*April to September*
General and Scientific Course	1	Lectures and lab work	'Workshop'
	2	Lectures and lab work	Pupilage
Technical Course	3	Lectures and lab work	Pupilage
	4	Lectures and lab work	Pupilage
Practice	5	Pupilage	Pupilage
	6	Pupilage	Pupilage

Dyer's programme explicitly warned against misconstruals of training in the college 'workshop', and reiterated the integrity of the authentic commercial workshop: 'It is not intended that the work done here shall be sufficient to train the students as mechanical engineers. For that purpose they must go as well to a regular engineering establishment.'[80] He further clarified the role of the physical laboratory that Ayrton was to run for third and fourth year students specializing in civil engineering, mechanical engineering, telegraphy and practical chemistry. Following the contemporary idea of physics laboratories as places to learn 'exact' modes of behaviour, Dyer emphasized that the 'special object' of the physical laboratory was to enable students to become 'practically acquainted with the fundamental laws of Physical Science, as a foundation for accurate reasoning upon physical phenomena, and the applications of the principles of physics to engineering'. Whilst the specific purpose of the laboratory for students of telegraphy was to teach 'the laws of electricity and magnetism, with their practical application to telegraphy', it was made explicit that during the summer pupilage they would be 'instructed in practical testing of lines and batteries and in the use of telegraphic instruments', i.e. this was a process of initiation which could not take place in the laboratory, only in the field of practice.[81]

Trade journals in England commented enthusiastically on Dyer and Ayrton's technical education scheme for the Japanese: the *Telegraphic Journal* in February 1875 publicized the 1874–5 *Calendar* of the 'remarkable college' and described the syllabus of the course of telegraph engineering as 'very complete'.[82] Although Japan had been connected to Britain by the ubiquitous telegraph cable since 1871,[83] its problematic cultural relationship with Dyer's missionary experiment in technical education was not as visible. Although Brock cites evidence from Ayrton that his first laboratory was crowded and unpleasant to work in, it would seem that the major problems faced by Ayrton lay not in the internal operation of the laboratory but rather in integrating the products of his teaching into Japanese society.

In his annual report to Dyer in October 1877, Ayrton discussed such problems. He wrote first that trainee telegraph engineers in the college laboratory were 'well-practised' in the construction and use of galvanometers, electrometers, resistance coils and condensers, etc. They were trained to carry out all the standard electrical, mechanical and chemical tests on all telegraph materials and to make joints in iron and insulated wire. They were well-practised in the laboratory 'performance' of all the tests employed in a land-line or submarine cable testing office, such as those used by Ayrton in India to detect and locate faults that interrupted continuous communication. For this purpose Ayrton had arranged for 'artifical lines' to be constructed out of a several hundred yard length of telegraph cable with sufficient resistances and condensers connected to simulate 'as far as practicable' the behaviour of a full-length cable—Jenkinian objections notwithstanding.[84]

The use *only* of artificial lines for laboratory testing was not to Ayrton's satisfaction. Not only did it fall short of Dyer's scheme for giving students authentic field practice in cable testing during the summer months of pupilage, it also meant that Ayrton could not communicate to his students the expertise he had acquired in India as an assistant to Schwendler. Ayrton had little choice in the matter since the Japanese Telegraph Department would only give the summer pupilage students positions on *cable-laying* expeditions, not *cable-testing* duties on the government's national network. Ayrton argued from the perspective of Dyer's curriculum that permission for testing should be granted,

> Since, not only would the students learn better how to cope with practical difficulties, but the results so obtained would be of great value to the Telegraph Department in the rapid determination of the position of interruptions, and in their speedy removal, as well as by the gradual improvement of all lines leaving Tokyo, through the regular elimination of the parts found bad by testing.

Although the Telegraph Department would not sanction the involvement of unqualified students in managing its cable network, it had made some concessions to the measurement expertise of Ayrton's regime: since the start of his course four years previously the Department had instituted 'a complete system of *quantitatively* testing new insulators' at the telegraph stores in Shiodome.[85]

Such concessions only raised further problems concerning the conflict between the European disciplinary culture of measurement practices, in which Ayrton's students were inculcated during their course of training at the College, and the native Japanese culture into which their skills had eventually to be reassimilated. Ayrton stressed that his students would ultimately reach a 'moderately high standard of excellence' in working for the Telegraph Department only if they received the 'most careful supervision of skilled foreigners'. In an imperialistic vein he explained that dire consequences resulted when his former students 'mixed only with the less intelligent Japanese': forgetting models of 'clever engineers, philosophers etc. in Europe and America [they] relapse into the orthodox small Japanese officer whose fancied knowledge is too vast to allow them to attend to trifles.' Ayrton argued that in telegraphy more than in any other science, 'success depends upon trifles: the difference of one hundredth of an inch in the position of a wire means the difference between . . . a telegraph line in good working order and a total interruption.' Ayrton concluded dolefully that the only way of teaching students to learn the requisite 'habits of responsibility which are at present quite unknown in this country' was to impose an elaborate system of 'foreign supervision' to which the Japanese were obviously hostile.[86]

As a result of these problems and the broader indigenous political unrest of 1877 alluded to by Brock, Ayrton wrote to England enquiring of his former mathematics teacher at University College, Thomas Archer Hirst, whether it was feasible to reimport his component of Dyer's scheme of technical education.[87] He received a negative response to his enquiry, but following the opening of a laboratory complex built to his specifications in 1877, Ayrton became occupied with the implementation of a new regime of disciplined measurement for his students. From an account of this well-ordered regime in an anonymous article entitled 'A visit to Professor Ayrton's laboratory', published in the *Japanese Weekly Mail* in October 1878, we can discern the devices and strategies deployed by Ayrton to maintain this discipline.[88]

The feature of Ayrton's laboratory complex most graphically highlighted in this account is one that shows how far he replicated the physical conditions of stability found in William Thomson's Glasgow laboratory.[89] The author of the article wrote that 'the great peculiarity of the whole laboratory seems to be, that nearly every table in the place rests on columns of masonry coming up from the foundations, and kept detached from the floors and walls, so that instruments resting on these tables may not be shaken by persons walking about in their neighbourhood.' A further device for stabilizing the working environment of this laboratory was also introduced by Ayrton: 'It is possible for students to leave his batteries and wires, his galvanometer, and little temporary contrivances, for by pulling down a glass door, he shuts out dust and meddling hands. That there is a great saving of time effected by these shutters must be evident.' Such was the compelling nature of Ayrton's regime and his 'fine' instrumentation that the students working in this laboratory had many times shown themselves to be 'eager' to work until 10 p.m.[90]

Next to this lecture theatre turned laboratory was another laboratory similarly equipped, with stone-supported tables shared between two or more students who were

> mainly telegraph students, testing lengths of wire and insulators and batteries, using Thomson's [mirror galvanometers], with their flashing spots of light, and peg-boxes, and batteries. The telegraph students were in this room where the vices and anvils, and tool-racks, and furnaces were; and some of them were busy making joints in telegraph wire, and they were all busy, and seemed to like their work and they were evidently all well up to their work. This is a model workroom, the tool-racks all in that state of perfect completeness which shows constant use and constant attention, the countless drawers each with its drawer alphabetically arranged, the cases of apparatus overhead, and the general atmosphere of efficiency would have tempted the laziest man into using files and hammers, and shellac.[91]

The feature of this account that we should focus on here—in order to see a contrast with Ayrton's later laboratories—is that the highly ordered and well-maintained technical regime in Ayrton's laboratory visibly engendered extremely ordered and efficient experimental conduct among his students. It is important to note that there were two independent criteria used by contemporaries to ascertain whether this laboratory was a success: the transmissibility of its products to a sympathetic external culture (on Ayrton's terms the laboratory was thus a failure) and public judgement upon the internal disciplinary operation of the laboratory (on the journalist's terms the laboratory was thus a success).

For further evidence of this interpretation of 'success' one can discern the impression made upon the author by the innermost sanctum of the laboratory, the 'dark chamber':

> In this sacred place there were six columns of stone, each standing about four feet above the floor. Two were devoted to a Thomson's electrometer and its scale. This wonderful instrument had been taken to pieces some five times since its arrival in Japan, and it says much for the patience of Mr Ayrton and his students that the instrument was in good order. It is probable that this is the only specimen of the instrument which has survived even once being taken to pieces in a physical laboratory.[92]

Finally, we can see how far this laboratory was seen as an indoor environment where *scaled down* models of telegraphy, *pace* Rankine, had a legitimate didactic function:

> It would take too long to describe fully the working of the telegraph museum, with its specimens of all kinds of telegraphy, posts, insulators, sending and receiving instruments etc. It is sufficient to say that every instrument was in working order, as one could tell by touching the various keys, when the clicking of the receiving instrument might be heard.[93]

Despite the spectacular facilities that this new laboratory gave to Ayrton for teaching his Japanese charges, things clearly did not turn out to his liking, probably for the political reasons mentioned above. As early as December 1877—within a few months of the opening of this laboratory—Ayrton clearly intended to return to England: he wrote to the Council of the STE, as Honorary Local Secretary for Japan, tendering his services in the post of the Society's London Secretaryship. This offer was repeated in a letter of 18 February 1878 in which Ayrton informed the STE that he would leave for England in June of that year, suggesting Perry as his replacement in the local Japanese position. By October 1878, within a year of his laboratory's opening, Ayrton returned to England, significantly *without* any clear prospect of a comparable academic post, to act instead as a consultant to the telegraph manufacturers Messrs Latimer Clark, Muirhead & Co. and to edit the *Journal* of the STE from 21 October.[94]

THE ELECTRICIAN AND LABORATORY EDUCATION FOR THE TELEGRAPHIST IN 1878

> As far as I am aware all that has been done in Great Britain in the [systematic] teaching of telegraphy may be summed up in the laboratory courses of, on the principles of electrical measurement, given separately by Professors Sir William Thomson and G.C. Foster to certain students of the Indian Government Telegraph Department, and the practical course of instruction given by Mr Preece at Southampton to the same students, but none of these courses lasted sufficiently long for anything like a complete technical education to be imparted, so that they had to be continued in the testing rooms, workshops and signal offices in India, and by learners being attached to all constructing parties in that country. . . . Now, although the facilities that are thus afforded to intelligent students to gain practical information by a daily study of the machinery and works are very great . . . still the principle that no regular instruction is imparted, and that the students are left to pick up what scraps of education they can, is likely to lead to very irregular attendance and very slow progress in the case of the less bright pupils; consequently such a system cannot be commended.[95]

Ayrton's return to England as a professional consultant for a telegraph company coincided with rapid and interrelated developments in electrotechnology and debates on technical education. During 1877, Alexander Graham Bell's and Thomas Alva Edison's rival versions of the telephone caused great excitement on both sides of the Atlantic when they were presented to the world as aural extensions of telegraphic science. Bell went to London in 1877 and at a 'crowded meeting' of the Physical Society in December he lectured upon and gave experimental illustrations of his instrument's capabilities, to a considerable response from J.H. Gladstone, Foster, Guthrie and others.[96] The *Telegraphic Journal* reported that its specialist interest had been in a somewhat depressed state during the course of that year, irrespective of extensions in duplex telegraphy and innovations in quadruplex systems, so the resources of electrical manufacturers had been diverted not only to producing telephones but also to electric bells and sewing-machine engines.[97]

At the end of 1878 the journal changed its title to become the *Telegraphic Journal and Electrical Review* in acknowledgement of the public debuts of David Hughes's microphone, Edison's phonograph, and the electric lighting systems of Jablochkoff, Wilde, Werdermann and Edison, powered by the magneto-electric machines of Gramme, Siemens, Brush and de Meritens. Particularly with regard to the Jablochkoff and Edison laboratory-gestated innovations, the journal confidently predicted the demise of gas lighting, and a 'permanent gain to the electrical profession' in the new employment opportunities for lighting engineers.[98] The subsequent complex history of rival inventors wrangling over electrical patents and the reform of the Society of Telegraph Engineers to recognize the new specialist 'electricians' who adopted the new technologies of lighting, power and telephony[99] is beyond the scope of this paper. However, analysis of the related debate on the professional training of these practitioners is essential in order to establish the context of 'technical education' in which Ayrton established his archetypal teaching laboratory at Cowper Street during early 1880.

In May 1878 *The Electrician* was jointly launched by Sir James Anderson, the former captain of the *Great Eastern*, on the Atlantic cable-laying expedition of 1866, and John Pender, telegraphic manufacturing magnate and Liberal Member of Parliament. From the outset they appointed an editor, C.H.W. Biggs, to use the periodical not only to discuss innovations produced by those 'steadily working in the laboratory'—citing Bell, Edison, Gladstone, Maxwell, Thomson, Varley and others—nor just to report meetings of the STE and Physical Society, but also actively to promote the proprietorial view of suitable training for an electrical practitioner by advising readers on technical education suitable for the telegraphist and electrician.[100] In the editorials written by Biggs around the time of Ayrton's return to England we can see how the latter's Japanese laboratory was canvassed by *The Electrician* as a model for such education in Britain.

In September 1878 Biggs criticized the system of training developed by the 'Electrician' of the Post Office, W.H. Preece, for the telegraphers under his jurisdiction, as one that undervalued scientific instruction in favour of mere 'manipulative skill' with instruments.[101] Later in the same month Biggs harangued the STE for making no effort to coordinate or legislate upon the proper scientific education of electricians despite there being no lack 'of really sound and deep science in a body that has . . . such members as Profs. Adams, Ayrton, Thomson etc.'. The many students who were unable to afford the university-level courses of Thomson, Adams and Foster, Biggs declared, could obtain the requisite instruction only from private schools or teachers, at considerable expense, or in the office of a company 'where the training cannot but be hasty and empirical'.[102] Following consultation with W. Lant Carpenter, an *Electrician* columnist who was also joint manager of the London School of Submarine and Military Technology, Biggs acknowledged in the next issue that this school had indeed offered 'a thorough practical training with sound theoretical instruction' since its foundation in 1868 and that its graduates largely constituted 'the rising generation of telegraphists and electricians'.[103]

Notwithstanding what Biggs described as this school's 'vast amount of unostentatious work in the direction of telegraphic education', he laid down an agenda for instruction which he considered no extant college course to have met: a comprehensive course in telegraphy that would also prepare electrical students to be employable in other technologies. The overall desideratum was a disciplined course of well-executed precision measurement experiments:

> For this course of instruction in the theory of electricity very delicate instruments . . . are absolutely requisite; and, what is of yet greater importance, the skill to use such instruments, the faculty of giving demonstrations which from their neatness and certainty are calculated to encourage the student in the use of methods of precision, should also be forthcoming.

Focusing more particularly on the case of telegraphy, the editor stressed the importance of some institutional surrogate of outdoor practice using authentic instruments under the supervision of a demonstrator with knowledge of professional work:

> In the first place, all the signalling instruments in ordinary use, such as the single needle, the Morse recorder, and Wheatstone's automatic recorder, require to be represented, duly arranged in a circuit, and to be taught by a preceptor who, in addition to manipulative skill, has the more rare talent of communicating to others his own knowledge, together with labour saving method and good 'style'. Then in the more advanced manipulation class, the learner has to make the acquaintance of the 'mirror receiver' and 'syphon recorder'. But in order that his experience of these instruments should be any of real utility, he must work with them under the conditions of practice, that is they must be circuited on a cable, real or 'artificial'.

Since 'real' cables were manifestly not feasible he recommended the use of Varley and Muirhead's expensive and thus rare 'artificial cable', which consisted of a long sequence of capacitors and resistors, as a *sine qua non* for 'an efficient school of telegraphy'.[104]

Lant Carpenter's partner in the running of the School of Submarine and Military Telegraphy, W.N. Tiddy, wrote to Biggs in October angrily asserting that all these desiderata were met by the school, which had 'for many years past' incorporated the testing of artificial cables, an activity for which they had received testimonials of approval from such authorities as Sir William Thomson FRS and Cromwell F. Varley, Esq., CE, FRS.[105] Biggs's propagandistic concern, however, was with systematic instruction for electrical students at all educational establishments. Hence in his next editorial he directed attention to the most comprehensive and 'advanced' course available in the country, at University College London, in which students received training in both G.C. Foster's physical laboratory and A.B.W. Kennedy's unique engineering laboratory.[106] His essential complaint was that the 'technical education' of telegraph engineers at UCL was not taught in a specialist laboratory under the attention of a single professor, but only in Foster's and Kennedy's laboratories, where areas of existing curricula could be negotiated to fit accordingly.[107]

In late October Biggs introduced to his readers the Imperial College of Engineering in Japan as 'the only college in which telegraphy is systematically taught in the English language', i.e. in a specialist laboratory under a professor who was also a practising telegraph engineer. In addition to the museums of working models in civil, mechanical and telegraph engineering, Biggs emphasized:

> Here, also, we have a laboratory with a workshop attached, a laboratory be it remembered, distinct from the physical and engineering laboratories, devoted solely to students of telegraphy, and in which we presume much good work must be performed, judging from the numerous results of original research that have issued from it [jointly by Ayrton and Perry] during the last few years, among which we may mention the measurements of the different specific inductive capacities of gases, contrary to the views of Faraday . . . some valuable contributions to Professor Clerk Maxwell's electromagnetic theory of light, and the recent re-determination in a new way of that important electrical constant *V* . . . not to mention a large number of purely telegraphic communications that have from time to time appeared in the various telegraphic periodicals of the day.[108]

The important message here was that Ayrton's telegraphic laboratory was an admirable archetype because it met Biggs's desiderata in being one in which high-calibre measurement experiments were carried out, and from which legitimate extensions of telegraphic science had emanated to benefit the world of practice.

A further important message came from Biggs's reading of Henry Dyer's syllabus for the Japanese College concerning the component of a telegraph engineer's training that lay beyond the territorial prerogative of the specialist laboratory:

> Some part of their knowledge of surveying and of construction . . . can be obtained from lectures and from experiments on strength of material etc. performed in the laboratory; but the actual practice—that practice which alone makes perfection—must necessarily be obtained in the field. We would, therefore, recommend some such scheme as that followed at the Imperial College of Engineering in Japan, where with the exception of the first year's course, half the year is spent at lectures and in the laboratory, and the other half surveying and constructing actual telegraph lines.[109]

This editorial was published, coincidentally, on the same day, 26 October 1878, as the article on the journalists' tour of Ayrton's laboratory in the *Japan Weekly Mail*. In the issue of 2 November Biggs published, intriguingly, an item that clearly came from a source close to Ayrton—if not from Ayrton himself, since he had now returned from England. This item was Ayrton's 'Report on the Course of Telegraphic Engineering, Imperial College of

Engineering, Tokei, 1st October 1877', originally sent to Henry Dyer as a formal private letter. The report, discussed above as evidence of Ayrton's cultural difficulties in Japan, was published in full and prefaced by the comments of Ayrton that form the epigraph to this section. It is important to note that Biggs gave space to an account by W.G. Adams of the physics laboratory-based course in telegraphy he ran at King's College, London—one obviously not familiar to Ayrton—which trained telegraphists in cable-testing to Biggs's specifications. Nevertheless, maintaining the line that the Japanese College was the prototype for British institutions, Biggs succinctly introduced the letter as 'information that will perhaps aid in the development of Telegraphic Education', embodying as it did all of Biggs's messages about the uniqueness of Ayrton's laboratory and its merits in offering extensive training in testing 'artificial cables', complemented by a very substantial amount of field-work in cable laying under a practising engineer.[110]

The debate in *The Electrician* on the propriety of existing physics laboratories versus Biggs's mooted telegraphic laboratories continued in the issue of 16 November. While acknowledging complaints from academic laboratory physicists who felt their teaching to be undervalued in his scheme of technical education, Biggs reiterated that a specific symbiosis was required between telegraphic laboratory and commercial practice. With a combination of these two a student 'should be so trained that he would not make the mistake into which the German Post-Master General recently fell of imagining that Bell's telephone could replace the instruments in ordinary use with telegraphic administration.'[111] Biggs's scathing dismissal of such new rival forms of electrical technology from across the Atlantic soon moved, however, to criticism of England's 'backwardness' in the field of discovery and invention. He cited Edison's ascription of American 'superiority' to the 'facilities afforded to experimentalists' in the USA and Bell's comment on the generally uneducated British mechanic as being able to work 'only in a groove'.[112]

At this time the whole tenor of broader British debates on technical education revolved around such issues as the anti-educational conservatism of the manufacturing public and the inaccessibility, indeed non-existence, of research and teaching institutions that would meet Edison's standards. These broader debates had moved on since their first urgent formulation in the aftermath of the 1867 Paris Exhibition; by November 1878 the situation had reached the point at which discussions between the British government, the ubiquitous T.H. Huxley and the vastly wealthy Livery Companies of London precipitated the foundation of the City and Guilds of London Institute for the Advancement of Technical Education (CGLIATE).[113] It was this philanthropic collective of London manufacturers that created the financial and administrative agency which William Ayrton deployed to create his unique laboratory courses in electrical engineering during the early 1880s.

THE CITY AND GUILDS INSTITUTE AT COWPER STREET 1879–1882: CREATING A SOCIAL SPACE FOR 'TECHNICAL PHYSICS' IN THE LONDON INDUSTRIAL CONTEXT

> [A]nother nation . . .—the Japanese—have set us an example which our ambition should lead us to emulate. . . . Ten years ago a feudal country . . . a nation whom we regarded as barbarous, that nation of whose manners we were comparatively ignorant, whose very modes of thought are so very different from our own we can be hardly said to understand them now. That people had but three years emerged from what was almost slavery, when up grew, in its very midst, a technical college, with its carefully chosen staff of English professors, with its laboratories, classrooms, museums, libraries, workshops . . . and to enter and study at this college, neither money, nor position, nor any qualification necessary but ability, and desire to study; so that working at the lathe, or conning over the books in the class-room, or experimenting in the laboratory may now be seen, side by side, the young nobleman and the young artisan.
>
> An example to emulate did I say? The City Companies have given us the opportunity, let us show that we appreciate it. They do not propose merely to teach each man his trade or business for that he can best learn in the workshop or counting house.[114]

To launch his career as City and Guilds Professor of Physics, Ayrton offered to his metropolitan constituency of electrical apprentices his Japanese laboratory as a model of well-financed, well-ordered and egalitarian technical education. P.J. Hartog wrote after Ayrton's death that his Japanese laboratory had indeed been a model for those which Ayrton later organized in England, 'and through them, for numerous other laboratories elsewhere'.[115] W.H. Brock has commented on this claim, arguing that Ayrton specifically imported Henry Dyer's workshop-laboratory curriculum into his City and Guilds teaching between 1879 and 1908.[116] In this respect it is important to appreciate how sensitively Ayrton integrated the operation of his electrical laboratories into the local milieu of industrial London, having learnt much from his lack of success in assimilating his Tokyo laboratory into local Japanese culture.

Ayrton's appointment in the summer of 1879 was well-received by the trade media, Biggs for *The Electrician* describing Ayrton and his colleague Armstrong as two of the 'ablest teachers' for the purposes of technical education.[117] Ayrton's inaugural address was published by both Biggs's journal and the *Telegraphic Journal and Electrical Review*, and contained messages of powerful contemporary currency. One message was that practical scientific education was a panacea for the nation's decline in European industrial prestige: it could serve the purpose of augmenting the innate superiority of the British worker's morals and dexterity, epitomized by Ayrton in the example of Michael Faraday. The other important message for his audience on 1 November was that Ayrton's lectures and laboratory teaching would not compromise the integrity of the workshop as the definitive place for the apprentice to learn his trade.[118]

Ayrton's commitment to this traditional integrity is discernible in his 'Report on the requirements of the Department of Physics at the proposed Technical Schools', presented to the CGLIATE on 31 December 1879.[119] He distinguished the mooted laboratory at the Cowper Street Schools from Continental 'trade schools', as a place where apprentices could learn skills *unobtainable* in the workshop. Ayrton declared, 'it is *only* in real workshops . . . that an apprentice can learn his trade; it is *only* where there is a fair proportion of workmen to apprentices that the latter can learn the dexterity of hand that characterizes the English workman.'[120] By commenting thus on the relative density of experts and unskilled novices in the workshop, Ayrton tacitly aligned himself with Fleeming Jenkin's position that an essential component in the skilled commercial use of tools or instruments could be learnt solely by the apprentice emulating the qualified engineer.

Actively dismissing the displacement of the workshop by the laboratory, he nevertheless argued for the incorporation of selected features of workshop technology and convention to confer commercial legitimacy on skills acquired in the laboratory of technical physics. He thus proposed that for a preparatory course of 'General Technical Physics', accompanied by lectures merely for 'guiding students' minds in their laboratory work' there should be 'a room in which students will perform general experiments in *Electricity*, measuring resistances, electromotive forces etc. with special reference to actual electrical instruments used in the arts'. For later specialist training a 'student of telegraphy would, for example, now work specially in the telegraphic laboratory'. He thus proposed the creation of a

> room in which students, who are following, or are about to follow, a trade intimately connected with *Electricity* would work. This room would contain in working order, the more important apparatus used in electrical engineering. Here students of telegraphy, electric lighting, electroplating etc. would see the immediate application of electrical and magnetic science to their daily occupations.[121]

Ayrton emphasized, however, that such work in a technical school would only serve to shorten apprenticeships from the conventional seven years to four if the teachers were themselves active professional practitioners, thereby conferring further workshop credentials on laboratory work.

On 2 March 1880 Ayrton reported to the CGLIATE that the laboratory courses in applied physics and chemistry had been 'taught with considerable success' in the Monday and Friday evening classes since November, not only because of the (apparently) abundant attendance, but also because Ayrton had acquired a permanent stable environment for his practical teaching. The Headmaster of the Cowper Street School, Dr Richard Wormell, had originally only lent Ayrton and Armstrong rooms after his pupils left at 4 p.m., but having 'fully realized the difficulty' of daily assembling and dismantling the equipment he sacrificed classroom space by furnishing Ayrton with two basement rooms and Armstrong with one.

While this facilitated the extension of teaching to part-time day classes, it rendered Ayrton's problems of acquiring authentic industrial equipment even more acute: 'I have had to resort to borrowing from instruments makers, manufacturers etc. a large quantity of apparatus.' Most would recall their equipment within a few days, thereby inhibiting the complete prosecution of an experiment.[122] Only with permanently installed equipment could Ayrton create the disciplined regime of measurement-oriented experiments that he considered a pedagogical prerequisite, and he therefore requested £200 for equipment such as mechanical models, pulleys, levers, valves, pumps, barometers, thermometers, chronographs and pressure measurement apparatus. He explained that with his two new laboratories and a part of the lecture theatre he could install 'a series of experiments all of which each group of students could perform *seriatim*; thus acquiring such a knowledge of applied physics as could not be obtained from any number of lectures or reading of text books.'[123]

Several evening students 'engaged in trades especially connected with electricity' had asked Ayrton to commence a 'special course of electrical engineering', to which request he nobly commented, 'I have already temporarily conceded, and these students have temporary laboratory work by themselves' in the Department of Applied Physics. 'In view of the fact that electrical engineering (including electric lighting, electric telegraphy and that most important subject the electrical transmission of power etc.) bids fair to become one of the foremost industries of the day', Ayrton suggested to the CGLIATE that they should 'avail themselves of his reputation as an electrician' to employ him at the professional rate after Easter to run such an unprecedented course, to which he predicted many new students would come. For such a special course on electrical engineering £120 would be the going rate, and 'for carrying on such a course there would be required certain apparatus of *permanent* value for electrical measurement, etc., necessitating a single expenditure of about £150':

Two sets of resistance boxes, 1–10,000 ohms with proportional coils for testing and for making artificial telegraph lines for students' experiments	... 50 0 0
High resistance reflecting Thomson's galvanometer	... 30 0 0
Morse instrument, with relay	... 36 0 0
Wheatstone's fast speed instruments, puncher, transmitter and receiver	about 45 0 0
Samples of insulators, wire, etc., for the students	... 5 0 0[124]

Here we see Ayrton attempting not only to recreate the equipment and conditions of his Japanese laboratories, but also to meet Biggs's agenda for a proper course of telegraphic education as laid down in *The Electrician*, particularly with regard to the artificial cable apparatus. It is interesting to note therefore the direct editorial response made to Ayrton's published reports by Biggs in May 1880. Citing substantial extracts from both reports discussed above, particularly the sections relating to telegraphic education, Biggs commented that 'the innovator who would do most good is he who accepts things pretty much as they exist, and attempts nothing more than raising his voice for something but slightly removed from that conceived in the popular mind.' So conservative was the Professor of Physics in his innovations that Biggs exclaimed, 'we are not sure if Professor Ayrton goes as far as we should go'. Noticing that 'Professor Ayrton is to give a special laboratory and tutorial course in electrical engineering', partly running during the daytime, Biggs suggested on his behalf that 'it seems important that employers should consider whether it would not be ultimately to their benefit to introduce some kind of half-time scheme amongst the apprentices.'[125]

By spring of the following year all of these guarded 'innovations' in technical education had been put into effect, both the *The Electrician* and the *Telegraphic Journal and Electrical Review* reporting Ayrton's 'special', i.e. daytime, syllabuses of 'Electrical Instrument Making' and 'The Electric Light'. Emphasizing 'as a matter of importance' that these courses were open to both male and female[126] students, both journals documented how beginners in the former course, for example, constructed electrometers of the Peltier and Thomson varieties and learned the cause of failure in (E.O.W. Whitehouse's) high-e.m.f. machines for telegraphy; in the advanced course students studied, for example, the construction of Thomson and Varley's resistance coils, condensers and their application to the construction and use of artifical cables; students of lighting dealt practically with electric arcs, lamps, candles and the magneto- and dynamo-electric machines of Gramme, Siemens and Wilde generally used to power them, in addition to some of Ayrton and Perry's proliferation of electromagnetic measurement instruments.[127]

In May 1881 Biggs once more editorialized approvingly on the City and Guilds Institute at Cowper Street as 'foremost' in the 'promotion of technical knowledge among the artisan classes', describing the syllabus as 'at once practical and attractive'. Biggs, now concerned with more than telegraphy, further opined upon the propriety of Ayrton's laboratories as being ideal vehicles for learning the finer rules of telephone construction: 'their every detail should demand the utmost care and nicety, so that the instrument may work up to its capacity'. He further praised Ayrton's arrangement of the lecture and laboratory hours for the lighting course to accommodate the late-afternoon working schedules of lighting engineers, and made comments verging on the complimentary regarding the introduction of women into the laboratory course.[128]

The increasing popularity of Ayrton's day and evening classes, due in at least some part to the canvassing of the two rival journals, was noted in flattering fashion in March 1882 by the *Telegraphic Journal and Electrical Review*. The editors Alabaster and Kempe congratulated Ayrton on his evening courses on electrical engineering, also emphasizing the telephone construction and artificial cable telegraphy, since 551 students had passed through the Cowper Street physics laboratories since October alone.[129] However, the crowding of the classrooms was becoming so serious that shortly afterwards Ayrton publicly apologized for this in a paper strategically presented jointly with John Perry, his recently appointed colleague in the new Department of Applied Mechanics, to the Society of Telegraph Engineers and Electricians, entitled 'Some remarks on the technical education of an electrical engineer'.[130]

In the context of the contemporaneous construction of the new City and Guilds College at Finsbury, Ayrton and Perry discussed their fact-finding tour of laboratories of physics and technical physics in German, Swiss and Austrian universities and polytechnics. They used the occasion to criticize what they saw as a lack of differentiation between the two species of laboratory, resulting from the non-practising status of Continental professors of technical physics—which crucially contrasted with their own *unique* status as teachers of technical physics who also practised as professional engineers. Ayrton and Perry argued that unless colleges of technical education could retain 'the services of qualified electrical engineers, it will not be electrical engineering that will be taught, but merely electricity, instruction in which can be obtained in nearly every college in the country; but which instruction alone . . . does not fit a man to become an electrical engineer'.[131] Explicitly advertising the nearly completed Finsbury College as an example of this British ideal, they apologetically alluded to the problems of the past few years' work:

> For the last two years such an education has been given, in a more or less hand to mouth fashion, in some small temporary rooms rented in Cowper St. The education has necessarily been fragmentary, from the absence of space, and especially from the absence of a sufficiently large teaching staff; but we venture to think that the men attending them have been able to gain much information of professional value to them. And we are led to this conclusion, not because we have had well polished cabinets full of highly lacquered instruments, but because some hundreds of men, after their day's work in electric light factories or instrument makers' workshops was over, have thought it worthwhile to come regularly, night after night, to learn how the efficiency of a electric lamp or of a dynamo is actually measured—how to obtain, experimentally, the characteristic curves of dynamo machines at different speeds, or to engage in the smaller experiments of calibrating galvanometers, testing magnets etc.[132]

Ayrton and Perry emphasized the disciplined system of rigidly fixed and ordered measurement experiments being carried out by groups of three students whose results were monitored for accuracy through the 'judicious use of squared paper'[133] as a prerequisite for such a professional education. Their concluding remarks pointed to the projected opening of the Finsbury College in the autumn of 1882 as the opportunity for this system of professional training to be exercised at its most efficient.

FROM CHAOS IN COWPER STREET TO TECHNOCRACY AT TABERNACLE ROW: THE PUBLIC SPECTACLE OF CREATING AN ORDERLY ELECTRICAL ENGINEERING LABORATORY

In their public pronouncement Ayrton and Perry once more alluded to the Japanese ideal of pedagogical order, which Perry had earlier advertised in January 1880 by publishing the account of Ayrton's orderly laboratory as reported in the *Japanese Weekly Mail* of 26 October 1878.[134] It is essential to appreciate that Ayrton achieved this pedagogical order in his London laboratories only after major public criticism of the apparent chaos that reigned under his management of the overcrowded and underfunded laboratories at Cowper Street between 1882 and 1883. In showing how Ayrton finally arrived at an orderly regime at Finsbury Technical College in 1884 I shall illustrate how, *pace* Brock, Ayrton's rigid pedagogical scheme was completely unlike his colleague H.E. Armstrong's famous 'heuristic method' of education-by-discovery.[135] I should also like to demonstrate that of the five features of the 'Finsbury method' pinpointed by Brock as common to the teaching of Ayrton, Armstrong and Perry,[136] the privileging of laboratory work over lectures was less conspicuous to Ayrton's industrial contemporaries as a characteristic of the Finsbury method than the *discipline* he eventually achieved in his experimental measurement courses.

Although in August 1882 *The Electrician* characteristically praised the Cowper Street Institute as being the (expected) source of the most successful students in the City and Guilds technological examinations in telegraphy,[137] by autumn the Finsbury building had not been completed. The subsequent even greater crowding in the Cowper Street laboratories led to a major source of embarrassing publicity for Ayrton in the pages of the *Electrical Review*. On 18 November 1882 Alabaster and Gatehouse invited electrical engineering students to comment upon the effectiveness of the teaching they received, clearly targeting Ayrton's classes at the Cowper Street Schools as a current subject of discontent.[138] The first reply they received was from 'F.W.F.' on 25 November:

> In the first place the rooms are too crowded; indeed so much so that I have frequently spent an hour in unsuccessful endeavours to perform an experiment. Secondly the only assistance the student receives, as a rule, in carrying out the experiments is from the written instructions attached to the instruments which are oft-times too vague to be understood by the beginner.

While praising the experiment-oriented delivery of Ayrton's lectures, F.W.F. explained that the severe overcrowding meant that the demonstrations were 'invisible to a large proportion of the students present'.[139]

On 2 December Alabaster and Kempe published the more damning criticisms of 'Seaboots': 'the experiments are *not* carried out under the direct supervision, either of the professor or of his assistants, students being left to find employment for themselves. The number of students is far in excess of the accommodation: one may see a dozen to twenty striving to read one am-meter and wondering what it is all about.' Seaboots, had, in fact, given up the Monday evening laboratory course altogether, 'for it is a mere waste of time when half-a-dozen bewildered students are each trying to measure the resistance of the same battery with the same apparatus at the same time, alarming results being the consequence'. Altogether, he argued, 'the classes are much to large to allow of individual attention, and such is not to be desired until the assistants are a wee more courteous'. In the same issue Alabaster and Kempe invited one such assistant, Ayrton's chief demonstrator Arthur C. Cockburn, to comment upon these reports of the laboratory regime breaking down into a riotous anarchy of failed or impossible experiments. Naturally Cockburn abruptly contradicted Seaboots's testimony, paternalistically emphasizing the *availability* of Ayrton and his three assistants, who were 'only too glad to smoothe away any difficulties the student may meet with'. While acknowledging that it was 'only too obvious to every one' that the laboratories were 'immensely overcrowded, in consequence of the unexpectedly rapid increase in the number of students attending the Electrical Engineering classes', he admonished Seaboots for his failures of patience and obedience to the pedagogical regime laid out for him. He concluded with a terse suggestion as to a course of action that Seaboots might take in order to ease the laboratory overcrowding.[140]

A letter from 'X' in the 9 December issue of the *Electrical Review* provides evidence that the public failure of Ayrton's laboratory courses of electrical engineering was received by more traditional workers in the industry as a vindication of traditional apprenticeship methods. X suggested that the disastrous failure of several electric light installations that he had recently observed was attributable to the incompetence of workmen which derived, in turn, from the want of 'sufficient practical instruction' in the Cowper Street courses they had attended. He submitted that 'it is a pity electric light companies will not forego some of the prejudice they appear to have against the employment of old tried telegraph engineers'. The following week an 'Old Subscriber' responded to the above attacks by requesting the editors to confirm that the standards of teaching at the main rival to Ayrton's City and Guilds Institute—the establishment run by Lant Carpenter and Tiddy since 1868, now retitled the School of Telegraphy and Electrical Engineering—were 'all that could be desired' for a prospective practitioner. Alabaster and Kempe declined to comment on this query.[141]

The editors' balanced construction of this debate can be seen from the manner in which they juxtaposed the views of X with a letter from Henry Sayers, a former student of Ayrton's, who defended his electrical engineering course on the basis of his experiences in 1880–1. Sayers testified that the laboratory apparatus for each experiment had then been 'carefully arranged', with comprehensive written instructions and ready assistance from Ayrton and his staff; he attributed the recent problems merely to overcrowding, sympathetically remarking that he 'should be very sorry to be one of the 300 students in the temporary classrooms and laboratories'.[142] However, 'One Who Wishes to Learn' argued the following week that whatever order there had been in the laboratory work of the 1880–1 session had disappeared by 1882–3, acidly remarking:

> Much depends upon the attention given by the Professor and the Demonstrators in the laboratory, but it seems to me the want of this, and the almost entire absence of organization and system, and not the overcrowding, are the cause of the admitted failure of the laboratory experiments, for I have noticed the same difficulties when only a small number of students happened to be present.[143]

Tellingly, Alabaster and Kempe allocated lengthy column space in the following issue to the views of J.B. Henck, a correspondent who had considerable experience of teaching physics in the USA using Edward Pickering's system.[144] Henck delicately legislated to both the Cowper Street instructors and their students what should be expected of both parties in an effective system of laboratory teaching: infallible apparatus, short, simple experiments, and Pickering's system of monitoring circuses of experiments by the use of wall-boards bearing cards displaying the identity of student(s) attempting each experiment. Finally, Henck argued that 'no laboratory instructor should have more than ten, or at most twelve, students to look after at one time; if he has it will not be fair to criticize his patience and efficiency'.[145] After this authoritative rejoinder to the criticisms of disaffected students, correspondence on the subject went into a lull from late December until shortly after the opening of Ayrton's new accommodation the following year.

Much ceremony attended the opening of the Finsbury Technical College (FTC) at Tabernacle Walk on 19 February 1883.[146] Among the festivities the Director of the College, Phillip Magnus,[147] was at great pains to establish in his opening address that the College was to be seen as a definitive solution to the vexed question of what should constitute 'technical education'. Congruent with Ayrton's previous pronouncements, Magnus declared that technical education should be a laboratory-centred training in the day or evening that would 'fit students for practical work in the factory or engineer's shop'. Explicating this with regard to what he expected to be the 'most attractive' of the four departments at Finsbury, Electrical Engineering (significantly altered from 'Applied Physics'), Magnus gave evidence from his own experience of Continental schools to support the view cited in the *The Times* that Britain 'could claim precedence over other nations in having established a good practical school for the training of workmen and foremen in electrical engineering'. As an oblique reference to the controversies discussed above, Magnus legitimated the status of the FTC as a 'good practical school' by referring to the strict regime of teaching: 'The instruction afforded is that of a college; the discipline that of a school. A definite course of instruction is laid down for each pupil.'[148]

Notwithstanding Magnus's discourse on such matters, Ayrton's courses were still severely criticized after the opening of the larger Finsbury laboratories, on 24 March 1883, the correspondent 'Light' writing to the *Electrical Review* with the 'very strong conviction that many others feel as I do', that

> after all the talk and opportunity for making improvement, things at Tabernacle Walk are now little better than in a state of chaos. In the laboratory are supposed to be sets consisting of all materials for carrying out certain experiments. Most of the apparatus is out of repair, not joined up or absent; and some five or six sets (out of what ought to be nearly twenty) only may be called complete. For making these experiments, as was pointed out in some previous

Figure 1 First year electrical engineering laboratory, Department of Physics, Finsbury Technical College, 1884. (*Source*: The Electrician, *4 July 1884*.)

> correspondence, the only instructions are some very indistinct ones . . . Which from usage have become tattered and dilapidated to an extent almost reaching uselessness.[149]

Light added further that the 'Chief Demonstrator', i.e. Cockburn, was principally occupied in taking tickets and names at the door, and switching on the electric light at the beginning of lectures. Light's sardonic message was essentially that the courses at Finsbury should not have begun until the building had actually been completed, and he was evidently not alone in this view.[150]

Biggs of *The Electrician*, while highly sympathetic to Ayrton's scheme of instruction in electrical engineering, commented in an editorial of 24 March 1883 that the operation of his laboratory was 'quite legitimately' the subject of such criticism, 'for it is only by the lessons of experience, and the sharp eyes of lookers-on that such a scheme can be moulded into perfect working order'. Matters improved over the course of the next year[151] and by July 1884 Ayrton and Magnus were in a position to distribute publicity material concerning 'the perfect working order' that the scheme had now, which subsequently appeared in both *The Electrician* and the *Electrical Review*.[152] The most explicit material is in *The Electrician* of 5 July 1884, in which Biggs incorporated a picture of the first-year electrical laboratory (Figure 1).

The laboratory shown was one on the first floor of the Finsbury building and shared important features with Ayrton's other laboratories, such as that dedicated to 'delicate measuring experiments in telegraphy etc.' on the ground floor. First, they were all symbolically illuminated by an Edisonian system of incandescent electric lighting powered by an Edisonian dynamo and all were strictly regimented in the technical and social arrangements of apparatus. As *The Electrician* put it:

> These rooms are arranged for the carrying on of an organised series of experiments in current and statical electricity. The peculiarity of the method adopted is that each experiment has all the apparatus required for performing it ready in position, together with printed instructions. The students work in groups of three, as shown in the engraving.

The Electrical Review explained further that:

> these groups [are] arranged so that students whose knowledge is about the same work together. The experiments are performed as nearly as possible in a specified order; and before a group of students is allowed to pass on to a new experiment, each member is required to show to the professor, or one of his demonstrators, his written out notes of the previous experiment, including any deductions he may have made from it. After any student has completed all these experiments in the regular course, he is set to carry out what may be called scientific commercial experiments—that is the kind of experiments a master of a works might arrange to have undertaken, to enable him, by the application of the principles of science to his trade, to turn out the article in the best and cheapest form.

This publicity material, as an accompaniment to the engraving issued by Magnus, served the purpose of publicly establishing that Ayrton's laboratory met Biggs's and Henck's criteria for a properly ordered system of technical physics. The rigidly mounted measurement experiments met Biggs's requirement for 'demonstrations which from their neatness and certainty are calculated to encourage the student in the use of methods of precision'. The demonstrator depicted in an authoritative position in the laboratory met Henck's principle that a demonstrator should have to superintend no more than ten students at one time, and Pickering's managerial wall-board is conspicuously displayed behind him to reinforce the impression of discipline held over the students.

Furthermore, this arrangement met Ayrton's and Magnus's desiderata for a laboratory course that would 'fit a student for the workshop'. It contained workshop electrotechnology: Edisonian lighting, tangent galvanometers, resistance boxes, ammeters, electrometers, Daniell cells etc. It complemented the functional industrial workshop by inculcating students, under the paternalistic regime of the demonstrator, in skills of commercially exact measurement free from the commercial and social constraints of workshop life.

CONCLUSION

> [In 1884] electrical engineering was so much in its infancy that it was possible to regard it as a branch of physics, as may be seen from the fact that some of the most important men in the Society of Telegraph Engineers were Professors of Physics. That Society is now the Institution of Electrical Engineers with some 2400 members, and the prominent men in it are no longer Professors of Physics but their colleagues—the Professors of Electrical Engineering. For during the last ten years departments of Electrical Engineering, distinct from the older Departments of Physics, have been established in practically all the chief colleges in Britain.[153]

In this paper we have seen how the physics laboratory was the venue for the pedagogical domestication of electrical technology. William Ayrton transformed this well-established domain for the practical teaching of electricity into a academic space that belonged distinctively to electrical engineering *qua* a subdiscipline of experimental physics. This transformation was legitimated by several factors: Ayrton's contemporaneous practice as a professional electrical engineer and his use of industrial electrotechnology conferred workshop credentials upon the operation of the laboratory; the instruction in his laboratory complemented rather than competed with the training of the workshop apprenticeship by endowing students with skills in exact measurement that were acquired—contrary to the traditionalist stance of his early associate Fleeming Jenkin—most efficiently in the academic environment by work on strategically pre-prepared experiments under the paternalistic supervision of the professor and demonstrators; the didactic regime he finally developed in 1884 was a well-ordered system that met the functional desiderata of internal and external critics, i.e. apprentice telegraphists and electrical engineers, academics, industrial technologists and trade media.

Clearly then the creation of a viable electrical engineering laboratory was a very public phenomenon, the stability and efficiency of which required the ratification of technogically interested social groups that resided beyond the laboratory walls. Yet as the quotation from Ayrton above illustrates, this electrical engineering laboratory cannot be contextually demarcated from its parent, the physics laboratory, without historical distortion. Indeed, the close connection between electrical engineering and physics was still very much apparent upon the centenary of Faraday's birth in 1891.

In that year the editor of *The Electrician* wrote, 'it is rather discomposing to have to associate the idea of centenary with the name of FARADAY. He and his achievements belong so intensely to our own day that it is a shock to begin to memorialize him to the past.'[154] Certainly manifestations of Faraday's currency as a public resource were very visible in, for example, the moves of the Hammond Electrical Engineering Company to rename its 'Electrical Standardizing, Testing and Training Institution' to 'Faraday House' precisely to coincide with these festivities.[155] No better example of irony could be found for the historian concerned with the invocation of Michael Faraday as the 'founding father' of electrical engineering: like many laboratory training courses in electrical engineering schools in the early 1890s, 'Faraday House' conventionally modelled its curricula on Ayrton's City and Guilds archetype. Such massed teaching classes of *commercially oriented exact electrical measurement*[156] could not be more strongly contrasted than against the unworldly Faraday's solitary and qualitative researches in the private institutional laboratory of Albemarle Street.[157]

Acknowledgements

I would like to thank Bruce Hunt, Frank James, Ben Marsden and Crosbie Smith for their advice and criticisms, and I gratefully acknowledge financial support from the British Academy and the Institution of Electrical and Electronic Engineers (New York).

Notes and References

1. British Association for the Advancement of Science, *Report*, 1898, (pt 2): 768.

2. 'Inaugural address', *Journal of the Institution of Electrical Engineers*, 1900, 30: 48; partially cited in J. Marsh, 'The role of the IEE in the advancement of technical education in Britain up to 1924', *IEE Proceedings*, 1988, 135A: 266.

3. *The Electrician*, 1890, 24: 318–19.

4. See Bruce Hunt's paper in this volume, and C.W. Smith & M.N. Wise, *Energy and Empire: A Biographical Study of Lord Kelvin*, (Cambridge, 1989), 446.

5. Op. cit. (3).

6. See R.A. Buchanan, *The Engineers: A History of the Engineering Profession in Britain from 1750–1914*, (London, 1989), 162–75. Buchanan discusses only the professional tension between works apprenticeship and education in *theoretical* science; he does not discuss the tensions between *laboratory* teaching and workshop training (see 172–3).

7. W.J. Reader (with R. Lawrence, S. Nemet and G. Tweedale), *History of the Institution of Electrical Engineers 1871–1971*, (London, 1987), 12–13.

8. Perry, op. cit. (2).

9. G. Gooday, 'Precision measurement and the genesis of physics teaching laboratories', *British Journal for the History of Science*, 1990, 23: 25–51.

10. D.W. Jordan, 'The cry for useless knowledge: education for a new Victorian technology', *IEE Proceedings*, 1985, 132A: 587–601.

11. W.H. Brock, 'The Japanese connection: engineering in Tokyo, London and Glasgow at the end of the nineteenth century', *British Journal for the History of Science*, 1981, 14: 227–43, and 'Building England's first technical college: the Laboratories of Finsbury Technical College, 1878–1926', in F.A.J.L. James (editor), *The Development of the Laboratory*, (London, 1989), 155–70. See bibliographical notes, 243, for further reference to Brock's published material.

12. Rollo Appleyard, *History of the Institution of Electrical Engineers 1871–1931*, (London, 1939); W.J. Reader, op. cit. (7).

13. This physicist turned electrical engineer, was Ayrton's successor at Finsbury Technical College in 1885—see J.S. Thompson and H.G. Thompson, *Silvanus Phillips Thompson: His Life and Letters*, (London, 1920), 124–51.

14. S.P. Thompson, *Michael Faraday, His Life and Work*, (London, 1898), viii.

15. Smith and Wise, op. cit. (4), 695 n., 745.

16. For discussion of the two major journals see P. Strange, 'Two electrical periodicals: *The Electrician* and *The Electrical Review* 1880–1890', *IEE Proceedings*, 1985, 132A: 574–81. Note that the *Telegraphic Journal* became the *Telegraphic Journal*

and Electrical Review in 1879 and formally became the *Electrical Review* in 1891.

17. Fleeming Jenkin, 'Presidential address to Section G of the BAAS', BAAS, *Report*, 1871, (part 2): 226.

18. G. Gooday, op. cit. (9), 32-6; Smith and Wise, op. cit. (4), 649-83; R. Sviedrys, 'The rise of physics laboratories in Britain', *Historical Studies in the Physical Sciences*, 1976, 7: 409-15, 418-20.

19. W.E. Ayrton, 'Kelvin in the sixties', *Popular Science Monthly*, 1908, 27: 267.

20. 'Art', *Telegraphic Journal*, 1876, 4: 17.

21. Gooday, op. cit. (9), 42; Sviedrys, op. cit. (18), 409-15, 418-20.

22. For details see W.E. Ayrton, 'On some points in connection with the Indian telegraphs,' *Journal of the Society of Telegraph Engineers*, 1873, 2: 180-99, [discussion 199-205].

23. 'William Edward Ayrton, F.R.S.', *The Electrician*, 1892, 28: 346; 'Obituary: William Edward Ayrton, F.R.S.', *The Electrician*, 1908, 62: 187; 'Prof. William Edward Ayrton, F.R.S.', *Nature*, 1908, 79: 74; 'Ayrton, William Edward (1847-1908), *DSB*, 'William Edward Ayrton 1847-1908', *Proceedings of the Royal Society*, 1911, 85. For allusion to his work under Preece at Southampton see *Telegraphic Journal*, 1875, 3: 25.

24. Ayrton, op. cit. (19), 264.

25. For biographical information see *DSB*, W.J. Millar (editor), *Miscellaneous Scientific Papers by W.J. Macquorn Rankine*, (London, 1887), Memoir by P.G. Tait, xix-xxxvi; and S. Colwin & J.A. Ewing (editors), *Fleeming Jenkin, 1833-1885, Papers Literary, Scientific etc.*, (London, 1887), memoir by Robert Louis Stevenson.

26. See Gooday, op. cit. (9), 49.

27. *Minutes of the Royal Commission on Scientific Instruction and the Advancement of Science*, 1870, question 1753.

28. Ibid., q. 1756.

29. Ibid., q. 1578.

30. Ibid., q. 1759.

31. Ibid., q. 1760.

32. For a sociological analysis of how the 'authenticity' of Victorian practical skill was cognitively tied to specific learning regimes institutionalized in specialized socio-technical domains see G. Gooday, 'Laboratories *vs.* workshops: measurement and morality in Victorian electro-technology', unpublished paper presented at conference *Rediscovering Skill*, Newton Park College, Bath, September 1990.

33. *Minutes of Royal Commission on Scientific Instruction and the Advancement of Science*, 1872: q.9511.

34. Loc. cit.

35. Gooday, op. cit. (9), 29.

36. Op. cit. (27), q. 1683.

37. Reader, op. cit. (7), 6; Gooday, op. cit. (9), 29.

38. Appleyard, op. cit. (12), Reader, op. cit. (7); for criticism of Russell Moseley's survey discussion of the Physical Society see below.

39. Reader, op. cit. (7), 28; Appleyard, op. cit. (12), 48.

40. See, for example, *Telegraphic Journal*, 1877, 5: 69-71.

41. The military men were Captain P.H. Colomb, Major R.H. Stotherd, Captain C.E. Webber, Captain E.D. Malcolm, Major F. Bolton; the practising civilians were Louis Loeffler and Robert Sabine (Charles Wheatstone's nephew), Appleyard, op. cit. (12), 29-31.

42. Smith and Wise, op. cit. (4) 652-722; Gooday, op. cit. (9), 32-4. See also D. de Cogan 'Dr E.O.W. Whitehouse and the 1858 trans-Atlantic Cable', *History of Technology*, 1985, 10: 1-15.

43. 'Obituary notice: Edward Orange Wildman Whitehouse', *The Electrician*, 1890, 24: 319.
44. Appleyard, op. cit. (12), 38.
45. Buchanan, op. cit. (6), 69–76.
46. Appleyard, op. cit. (12), 42.
47. 'Rules and Regulations', *Journal of the Society of Telegraph Engineers*, 1872–3, 1: 10.
48. Reader, op. cit. (7), 29.
49. Appleyard, op. cit. (12), 34.
50. Appleyard, op. cit. (12), 34–5.
51. 'List of Officers, Members, and Associates', *Journal of the Society of Telegraph Engineers*, 1872–3, 1: 1–9.
52. Reader, op. cit. (7), 29.
53. See Gooday, op. cit. (9).
54. Ayrton, op. cit. (22) 180–99, [discussion 199–205]. For his earlier publications on testing techniques and technology see Ayrton, 'On a quantitative method of testing a telegraph earth', *Asiatic Society Proceedings (Bengal)*, 1871, 40 (pt 2): 177–85; 'On a form of Galvanometer suitable for the quantitative measurement of the electromotive force and internal resistance of telegraph batteries', *ASP (Bengal)*, 1871, 40: 217–21.
55. G.C. Foster, 'On a modified form of "Wheatstone's bridge", and methods of measuring small resistances'; *JSTE*, 1872–3, 1: 196–206; W.G. Adams 'On Latimer Clark's method of measuring differences of electric potential', *JSTE*, 1874, 3, 86–92.
56. Appleyard, op. cit. (12), 47; *Minutes of Council Meetings, Society of Telegraph Engineers*, 1874, 1: 149–50 (28 April).
57. See also R. Moseley, 'Tadpoles and frogs: some aspects of the professionalization of British physics, 1870–1939, *Social Studies of Science*, 1977, 7: 426, for the suggestion that the Physical Society was created by Guthrie in sympathy with a BAAS Report of 1874 which recommended the promotion of practical physics teaching in schools.
58. For more details see Chapter 8 of G.J.N. Gooday, *Precision Measurement and the Genesis of Physics Teaching Laboratories in Victorian Britain*, (unpublished PhD thesis, 1989), University of Kent at Canterbury.
59. For example, F. Guthrie, 'On approach caused by vibration', *Proc. R. Soc.*, 1870, 18:93–4.
60. Jenkin to Stokes, 27 May 1873, Royal Society Archives. R.R.7.244.
61. Maxwell to Stokes, undated (1873), Royal Society Archives. R.R.7.245.
62. *The Physical Society of London 1874–1924, Proceedings of the Jubilee Meetings, March 20–22nd, Special Number*, (London, 1924), 20.
63. F. Guthrie 'On a new relation between heat and electricity', *Proc. R. Soc.*, 1873, 21: 168–9.
64. Op. cit. (62), 17. A primary source supporting Fleming's interpretation can be found in the Society's first Report, *Proceedings of the Physical Society*, 1875, 1: 5.
65. The circular issued by Guthrie is reproduced in G.C. Foster's obituary notice, 'Professor Frederick Guthrie', *Proceedings of the Physical Society*, 1887, 8: 9.
66. Maxwell wrote to Adams in December 1873 to say that he did not 'approve' of the Society's goals, explaining significantly that if the society was to 'publish Papers on Physical subjects which would not find their place in the transactions of existing societies or in scientific journals, I think its progress towards dissolution will be very rapid.' Maxwell to Adams December 1873, cited in L. Campbell and W. Garnett *The Life of James Clerk Maxwell*, 1882.

67. *Nature*, 1873, 9: 113. Note also that Lord Rayleigh joined the Society in 1877 and that two young researchers at the Cavendish laboratory, R.T. Glazebrook and J.H. Poynting joined the day after Rayleigh's appointment as Maxwell's successor in 1879, see Gooday, op. cit. (58).

68. Op. cit. (13), 20.

69. Foster, op. cit. (65), 9–10.

70. *Nature*, 1873, 9: 113; *Jubilee Proceedings*, 18–19, 22.

71. J.A. Fleming, 'On the contact theory of the cell', *Proceedings of the Physical Society*, 1874, 1: 11.

72. Moseley, op. cit. (57), 426–7, cites the comment of Shelford Bidwell, 1898 President, that in the early years of the Physical Society papers presented were 'sometimes blemished by serious errors'; Moseley, however, disingenuously omits to record Bidwell's subsequent remark that 'the demolition of the authors added much to the interest and liveliness of the discussions', *Proceedings of the Physical Society*, 1898, 16: 12.

73. See *Proceedings of the Physical Society*, 1874–92, 1–8.

74. Similarly S.P. Thompson and John Perry were Presidents of the STE respectively in 1899 and 1900, and Presidents of the Physical Society in 1901 and 1906, Appleyard, op. cit. (12), 287–92.

75. Ayrton's appointment was announced publicly in April 1873; see 'An Engineering College in Japan', *Nature*, 1873, 7: 430.

76. 'A visit to Professor Ayrton's laboratory', *Japanese Weekly Mail*, 26 October 1878, cited in J. Perry, 'The teaching of technical physics', *Journal of the Society of Arts*, 1880, 28: 171–2.

77. Remark on Ayrton's laboratory in J. Perry, 'Obituary: William Edward Ayrton, F.R.S.', *The Electrician*, 1908, 62: 187.

78. Brock, op. cit. (11), 'The Japanese connection', 230–6.

79. Ibid., 231.

80. Reprinted in 'On the present education system of Japan', *British Parliamentary Papers*, 1874, 65: 53–81.

81. Ibid.

82. 'Japan. The Imperial College of Engineering, Tokei', *Telegraphic Journal*, 1875, 3: 25–6.

83. Appleyard, op. cit. (12), 24.

84. 'Report on the Course of Telegraphic Engineering, Imperial College of Engineering, Tokei, 1st October 1877', cited in 'Telegraphic education', *The Electrician*, 1878, 1: 284.

85. Ibid.

86. Ibid., 286.

87. Brock, 'The Japanese connection', (11), 236.

88. Cited in Perry, op. cit. (76), 171–2. For reasons dicusssed below it is almost certain that the anonymous author of this article made his trip to the laboratory much earlier than October in that year.

89. See J. Bottomley, 'Physical Science in Glasgow University', *Nature*, 1872, 6: 29–32, and Gooday, op. cit. (9), 40.

90. Op. cit. (76), 171.

91. Ibid.

92. Ibid.

93. Ibid., 172.

94. Correspondence cited in Appleyard, op. cit. (12), 50; *The Electrician*, 1892, 28: 346.

95. William Ayrton, quoted in op. cit. (84), 284.

96. 'Proceedings of societies—the Physical Society', *Telegraphic Journal*, 1877, 5: 316-17.

97. '1877', *Telegraphic Journal*, 1878, 6: 1-3.

98. '1878', *Telegraphic Journal and Electrical Review*, 1879, 7: 1-3.

99. See, for example, the critical editorials entitled 'The Society of Telegraph Engineers', in *The Electrician*, 1878, 1: 210-11, and *Telegraphic Journal and Electrical Review*, 1879, 7: 125-6. The STE officially became the 'Society of Telegraph Engineers and Electricians' in December 1880, Appleyard, op. cit. (12), 64-5.

100. 'To our readers', *The Electrician*, 1878, 1: 6; 'John Pender, Esq, M.P.', *The Electrician*, 1883, 11: 53. Strange, op. cit. (6), 574-81. The *Telegraphic Journal* was also concerned with educational matters—see 'The submarine telegraph service', *Telegraphic Journal*, 1877, 5: 169-70.

101. 'Government schools of telegraphy', *The Electrician*, 1878, 1: 185. Note that William Ayrton was a trainee under the aegis of Preece in 1868.

102. 'The Society of Telegraph Engineers', *The Electrician*, 1878, 1: 210.

103. 'Telegraphic education', *The Electrician*, 1878, 1: 235.

104. Ibid., 235-6.

105. W.N. Tiddy to Biggs, 9 October 1878, *The Electrician*, 1878, 1: 250.

106. See Gooday, op. cit. (9), for a discussion of Kennedy's laboratory.

107. 'The technical education of telegraph engineers', *The Electrician*, 1878, 1: 270-1.

108. Ibid., 271.

109. Ibid.

110. 'Telegraphic education', *The Electrician*, 1878, 1: 284.

111. 'The technical education of telegraph engineers', *The Electrician*, 1878, 1: 306.

112. 'Electricians and their employers', *The Electrician*, 1878, 2: 6.

113. J. Lang, *City and Guilds of London Institute: Centenary 1878-1978*, (London, 1978), 11-18; C. Bibby, *T.H. Huxley, Scientist, Educator and Humanist*, (London, 1959), 123-43.

114. William Ayrton, 'The improvements science can effect in our trades, and in the condition of our workmen', inaugural address, City and Guilds of London Institute, 1 November 1879. Reproduced in *Telegraphic Journal and Electrical Review*, 1879, 7: 370, and in *The Electrician*, 1879, 3: 299, 310-12.

115. *DNB*, (1901-13).

116. Brock, op. cit. (11), 'The Japanese connection', 239.

117. 'Technical Examinations', *The Electrician*, 1879, 3: 270-1.

118. Op. cit. (114).

119. *CGLIATE Reports to the Sub-committee on Cowper St Schools . . . for meeting at the Drapers Hall December 31st 1879*, Ayrton Collection, Imperial College Archives, FA1 1064, 21-31.

120. Ibid., 23. Ayrton's emphasis.

121. Ibid., 25-7. This corresponds to Brock's description of the 'analytic' (*vis-à-vis* synthetic) element of the 'Finsbury method'.

122. *CGLIATE Reports presented by Dr. Armstrong and Prof. Ayrton on extending science classes* [to executive committee for meeting of 8 April 1880]. Ayrton Collection 3542/1056, Imperial College Archives, 4-7.

123. Ibid., 5.

124. Ibid., 7.

125. 'The City and Guilds of London Institute', *The Electrician*, 1880, 4: 294-5.

126. For an account of Ayrton's involvement in the women's movement and the research on the electric arc carried out by one of Ayrton's earliest female

students, who later became his second wife, see E. Sharp, *Hertha Ayrton: A Memoir*, (London, 1926), 117–206.

127. *Telegraphic Journal and Electrical Review*, 1881, 9: 171–2.

128. 'Technical education', *The Electrician*, 1881, 6: 308.

129. 'Technical education', *Telegraphic Journal and Electrical Review*, 1882, 10: 189; also 'The City and Guilds of London Institute', *Telegraphic Journal and Electrical Review*, 1881, 9: 333–4.

130. W.E. Ayrton and J. Perry, 'Some remarks on the technical education of an electrical engineer', *Journal of the Society of Telegraph Engineers and Electricians*, 1882, 11: 389–98.

131. Ibid., 396.

132. Ibid., 396–7.

133. See W.H. Brock & M.H. Price, 'Squared paper in the nineteenth century: instrument of science and engineering, and symbol of reform in mathematical education', *Educational Studies in Mathematics*, 1980, 11: 365–81.

134. J. Perry, 'The teaching of technical physics', *Journal of the Society of Arts*, 1880, 28: 167–76.

135. W.H. Brock, 'Observe, experiment and conclude. Finsbury College's new course of experimental philosophy in 1879–1880', *History of Education Society Bulletin*, 1980, 26: 45–50; also, Brock, 'The Japanese connection', (11), 237.

136. Brock, 'Building England's first technical college', (11), 166–7.

137. 'City and Guilds Institute', *The Electrician*, 1882, 9: 300–1.

138. 'Notice to students (electrical engineering)', *Telegraphic Journal and Electrical Review*, 1882, 11: 393.

139. *The Electrical Review*, 1882, 11: 416.

140. *The Electrical Review*, 1882, 11: 437.

141. *The Electrical Review*, 1882, 11: 479.

142. *The Electrician*, 1882, 11: 454.

143. *The Electrician*, 1882, 11: 479.

144. See E. Pickering, *Elements of Physical Manipulation*, (New York, 1873–6, and London, 1874–6).

145. *The Electrical Review*, 1882, 11: 499-500.

146. 'New technical college at Finsbury', *The Electrician*, 1883, 10: 341.

147. See F. Foden, *Philip Magnus: Victorian Educational Pioneer*, (London, 1970), 119–93, for discussion of Magnus's work for that CGLIATE.

148. P. Magnus, *Technical Instruction*, (London, 1883), 3–32.

149. *The Electrical Review*, 1883, 12: 248.

150. See correspondence in *The Electrical Review*, 1883, 12: 269.

151. See positive remarks made in articles entitled 'The City and Guilds of London Institute', *The Electrical Review*, 1883, 13: 469 and 1884, 14: 292.

152. *Electrical Review*, 1884, 15: 16; *The Electrician*, 1884, 13: 181–2.

153. W.E. Ayrton, letter to the Secretary of CGLIATE, June 1894. MSS 21, 868/10, undated, 2–3, City and Guilds Institute Archive, Guildhall Library, London.

154. 'The Faraday Centenary', *The Electrician*, 1891, 27: 186.

155. F.W. Lipscomb, *The Wise Men of the Wires: The History of Faraday House*, (London, 1973), 10–19.

156. See Perry quote op. cit. (2).

157. See D. Gooding, ' "In Nature's School": Faraday as an experimentalist', in D. Gooding and F.A.J.L. James (editors), *Faraday Rediscovered: Essays on the Life and Work of Michael Faraday, 1791–1867*, (Basingstoke, 1985), 105–135.

'The Engineer Must Be a Scientific Man'

The Origins of the Society of Telegraph Engineers

† W.J. READER

In the nineteenth century the Institution of Civil Engineers was the Great Mother of all engineers. Some of the children were fractious, especially the Mechanicals, but then they were nasty rough boys, not really fit for the society of gentlemen, as the Civils unquestionably were. The Telegraph Engineers, on the other hand, were always very respectful. Their founders, several of whom were distinguished members of the Institution of Civil Engineers, referred to Mother as 'the parent Society', and for the better part of forty years used her hall for their meetings. However, the telegraph engineer was not the same kind of engineer as the civils or the mechanicals. He[1] did not build great public works, nor did he build steam engines. He dealt with the most important commercial application of electric current: the electric telegraph.

Until the mid-1830s electricity was no more than one of the exciting toys of natural philosophy upon which, in the Royal Institution, Michael Faraday was conducting interesting experiments. But the engineer was not a natural philosopher (to use the terminology of the time). He was a severely practical man, suspicious of all philosophical speculation, and to him electricity was of no consequence, for he could make no use of it. It was a strange force which nobody really understood, and it had little connection with the world of practicality.

Cooke and Wheatstone changed all that. They were an unlikely combination—Cooke an Indian Army officer turned medical student, Wheatstone a brilliant natural philosopher with a practical turn of mind —but they succeeded not only in developing a commercially viable system of electric telegraphy but also in persuading the railways to take it up. In the 1840s the railways grew very quickly, and the telegraphs with them.

The 1840s—the 'hungry forties'—have a bad name socially, and the respectable classes feared revolution, probably not without reason, but technically they were a marvellous decade, and among the marvels the electric telegraph was one of the greatest.

> There are periods in the history of all the sciences [says a writer in 1843] when truths, hitherto known and studied only by philosophers, become the property of the world at large, by being rendered applicable to directly useful purposes. . . . The useful applications of *electricity* have been so recent, and the strides now making so gigantic, that there has not been time to classify and illustrate them. One week's discoveries may greatly surpass those of the previous week, and the mind is almost bewildered by the various paths in which art is receiving contributions from science.[2]

In 1846, only nine years after Cooke and Wheatstone had gone into business, their patents were sold to the Electric Telegraph Company, incorporated by Act of Parliament on 18 June 1846 as the first large British undertaking in the newest of all industries: the electrical industry. Other flotations and amalgamations of telegraph companies followed and the railways went on expanding their systems. By 1868, in the United Kingdom, there were 4,119 telegraph offices and 22,036 miles of line, and the greatest expansion, after nationalization, was still to come. Telegraphy by that time had spread all over the world and a submarine cable, representing the most adventurous and glamorous branch of telegraphy, had been laid between Ireland and North America (Newfoundland).[3]

In 1872 William Siemens said, in an ornate burst of Victorian self-confidence:

> The great network of international telegraphy extends already to every portion of the civilised and semi-civilised world; it traverses deserts and mountain chains, it passes over the deep plateau of the Atlantic and over the more dangerous bottom of tropical seas; what would be good practice in one country or under one order of climatic influences would be objectionable, insufficient, or wholly impracticable under another; but all these systems are intimately linked together, and the knowledge of the Telegraph Engineer must apply equally to all.[4]

'The knowledge of the Telegraph Engineer' was a new kind of knowledge in engineering. Civil and mechanical engineers worked largely by the light of experience and were sceptical, as a rule, of theory and 'book learning'. So were some telegraph engineers, but none could afford to ignore scientific principles entirely. They were the essential foundation of the electric telegraph, which must have been one of the earliest major inventions to have arisen directly from 'pure science'. Moreover, knowledge of electricity, rightly regarded as a most mysterious phenomenon, bordering on the supernatural, could only be extended by experiment and calculation.[5] The telegraph engineer was the first kind of engineer who, to do his job properly, had to share the outlook of the scientist as well as being a competent technologist. 'The engineer,' said Sir William Preece at the turn of the century, 'must be a scientific man.'[6] The telegraph engineers showed the way.

By 1870 the telegraph engineers were conscious enough of their own identity to wish to found their own professional association, which they called the Society of Telegraph Engineers (STE). One of the founders, Captain Charles Webber RE, said the idea came to him 'one broiling afternoon in June' when he was in command of the Twenty-second Company, Royal Engineers, which was repairing a telegraph route along the Uxbridge-Oxford road for the newly formed Postal Telegraph Department. The men were resting, and their officer, sitting under a tree and chatting to the Post Office Divisional Engineer, mentioned that he had been to a meeting of the Association of Gas Managers and Engineers and suggested 'that a somewhat similar society was wanted to bring telegraph engineers together'.[7] It is a pleasant picture, only slightly dented in its credibility by the fact that about five years earlier Webber had told a version of the same story in which the Gas Managers figured but not the 'broiling afternoon', the shady tree or the Divisional Engineer,[8] which simply goes to show how dubious a source reminiscences are for the historian. There is no doubt, however, that during 1870 the idea of founding a professional association was being widely canvassed among telegraph engineers and that Webber was among the most energetic canvassers.

Professional associations are founded for a variety of reasons. The practitioners of some occupations, unsure of their status, wish to lay claim to the prestige which membership of an association may confer, especially if it has a royal charter. Others feel that malpractice needs suppressing and proper professional ethics need defining and encouraging. Others wish to establish minimum levels of examinable competence. Some, no doubt, are attracted by the idea of job protection, though they are unlikely to be explicit about it.

None of these reasons can be detected in what is known of the motives of the founders of the STE. They seem to have felt no misgivings at all about telegraph engineers' professional or social standing. On the contrary, they knew they were at the leading edge of the high technology of the day—a parallel with today's workers in information technology suggests itself—and that they lacked nothing in public esteem. They said nothing about malpractice or professional ethics and one assumes that malpractice was unknown (telegraph engineers, unlike members of other professions, had not the temptations arising from dealing directly with the public) and that a decent level of ethics could be taken for granted. Thus they saw no need for the society to set standards of qualification for entry into the profession; on the other hand, they were particular about qualifications (as discussed below) for membership of the society. If job protection occurred to any of them, they said nothing of it, even obliquely.

What they were very conscious of was their standing as men of science. Moreover, they had identified a gap in the ranks of the London learned societies. There was no society specializing in electrical science. The Electrical Society of London had failed miserably between 1837 and 1843 and nothing had taken its place. There was a Chemical Society, a Geological Society, a Mathematical Society, a Royal Astronomical Society, a Royal Botanical Society, a Statistical Society, a Zoological Society, and many other societies, more or less learned, all meeting in London during the crowded weeks of 'the season' between November and June, but there was no Electrical Society. It was that, rather than a professional association, that the Society of Telegraph Engineers was set up to provide.

The founders' ambitions were worldwide, as befitted the founders of the world's first telegraph engineers' society and the leaders of a profession in which Great Britain led the world. They intended to found an international society publishing its proceedings, perhaps, in three languages—'every educated person throughout the civilised world,' said William Siemens, 'speaks either French, German, or English'[9]—and a class of Foreign Members was set up, paying £1 subscription instead of two guineas (£2.10) in consideration of not often being able to come to meetings. William Siemens, attending an international conference on telegraphy in Rome late in 1871, wrote letters in French to various high dignitaries who all replied in French, except his brother Werner, who replied in German. William was highly gratified with the response, no doubt the more so because several of his recruits—not, apparently, including his brother—enclosed twenty-five francs for their first subscription.[10] In spite of this foray, there was no doubt where the STE's centre of activity would lie. 'London,' said Siemens, 'is unquestionably the proper seat for such a Society, because it is the principal centre of the Telegraphic enterprise of the world, and musters consequently the greatest number of Telegraph Engineers. It is a remarkable fact that the manufacture of insulated wire, and of submarine cables, is almost entirely confined to the banks of the Thames.'[11]

The Society's 'Objects', set out admirably concisely in the first Rules and Regulations, make this purpose very clear. 'The Society of Telegraph Engineers,' they say, 'is established for the general advancement of Electrical and Telegraphic Science, and more particularly for facilitating the exchange of information and ideas among its Members.'[12]

At the Society's first Ordinary General Meeting, on 28 February 1871, the point was elaborated by the first President, William Siemens. 'There is hardly a problem in electrical science,' he said, 'that is not of practical interest to the Telegraph Engineer; and, considering that electricity is not represented at present by a separate learned society, ranking with the Chemical or Astronomical Societies, I am of opinion that we should not exclude from our subjects questions of purely electrical science.'[13] Latimer Clark, one of the senior telegraph engineers present, said he hoped that the Society would develop into 'something more than a mere Telegraph Society' and 'gather into the compass of its proceedings a large share of the many valuable and interesting papers which we now see read before other Societies, and scattered among other scientific publications. When that time arrives, and not till then, I shall feel that the Society of Telegraph Engineers has assumed its proper function and position in the scientific world.'[14] Cromwell Varley, another leading figure in the telegraph world, agreed. It had been the opinion of many, he said, 'and I myself entertained the same opinion, that telegraphy alone was a subject scarcely large enough to keep a society like this on its legs. . . . This society, I assume, will gradually, by natural selection'—presumably Varley was a Darwinian—'develop more into an electrical society than into a society of telegraphy proper.'[15]

The original composition of the STE was very carefully controlled, both in the activities which it covered and in the eminence of the personalities persuaded to join it, to further the Society's main purpose. Major Bolton, the Honorary Secretary, and the small group closely associated with him, must have been very busy, and very persuasive, during the latter months of 1870. It cannot have been easy to induce a figure so distinguished in the scientific world as Sir William Thomson, later Lord Kelvin,[16] or so highly placed in society as Lord Lindsay, heir to the Earldom of Crawford, to lend their names to such an enterprise as Bolton and his friends were launching.

> The support of the two great Telegraph Administrations of Great Britain and India [said Siemens] is secured to us through the accession of the Directors-General and the Chief Engineers connected with those systems. The military branch of Telegraph Engineering is very fully represented by the distinguished chemist [(Sir) Frederic Abel (1827–1902)] and engineer officers charged with those departments. As regards professional Telegraph Engineers the list includes an array of names many of which will ever remain associated with important improvements and generally with early telegraphic enterprise by sea and by land.[17]

This was a reminder that both Cooke and Wheatstone were still alive and active and had been enrolled as members.

Science in the universities, with some exceptions, was not strong. But one statistic will perhaps demonstrate the STE's success in attracting scientific talent. In 1872 twelve of the 140 members—9%—were Fellows of the Royal Society, and the number, though not the proportion of the total membership, increased over the years. In 1911 there were thirty-six Fellows.

Another group whom the founders of the STC were most anxious to attract were the wealthy amateurs who in the 1870s still conducted valuable research. They particularly wished to bring in those whom Latimer Clark called 'that large body of private scientific workers who love and pursue the science of Electricity without any thought of regarding it as a profession.'[18] Lord Lindsay, later Twenty-fifth Earl of Crawford, was one of these. A little later Sir David Salomons, nephew of the first Jewish Lord Mayor of London, was another. Undoubtedly it was expected of wealthy men like these, as it was expected of wealthy Fellows of the unreformed Royal Society before 1847, that they would support the Society and its activities financially. Lord Lindsay obliged by allowing a *conversazione* to be held, no doubt at his expense, in his laboratory in Greek Street on 6 June 1872, when Schutz-Wilson, the STE's first paid Secretary, perpetrated an 'illuminated motto':

The Lords of Lightning we; by land or wave,
The mystic agent serves us as our slave.[19]

Sir David Salomons, who had a private workshop at his house at Tunbridge Wells, was also generous with hospitality, financial advice and money.

The STE's requirements for full membership for professional telegraph engineers were strict. They demanded that a candidate should have been 'regularly educated as a Telegraph Engineer, according to the usual routine of pupilage', and that he should have had 'subsequent employment for at least five years in responsible situations'. Alternatively he would be accepted if he had practised as a telegraph engineer on his own account for at least two years and had acquired 'a degree of eminence'.[20] In their insistence on the importance of practical experience and personal responsibility the present-day entry regulations of the Institution of Electrical Engineers (which the STE was to become) remain, in spirit, much the same, but 'the usual routine of pupilage' today means a university degree. No one would have dreamt of that in 1872 or for a great many years afterwards.

Alongside these regulations for the professionals, a gate was left open, or at any rate very loosely latched, for scientists. The Associate grade, although professionals were obliged to pass through it, seems to have been especially designed for the non-professionals. 'This class,' says the regulation, 'shall include persons whose pursuits constitute branches of Electrical Engineering'—a very early use of the term—'who are not necessarily Telegraph Engineers by profession, but who are, by their connection with Science, qualified to concur[21] with Telegraph Engineers in the advancement of professional knowledge.'

In spite of these arrangements, the STE did not attract enough scientists, as distinct from telegraph engineers, to satisfy its founders, and the Council was badly frightened when the Physical Society was founded in 1874. Bolton wanted an amalgamation 'to prevent their taking away our best members',[22] and no doubt he and others also feared that the newcomer would occupy the vacant ground in electrical society which the STE coveted. Latimer Clark joined the Council of the Physical Society in 1875,[23] a *modus vivendi* seems to have been arrived at, and the threat passed. Anxiety remained, and in 1879 Latimer Clark wrote of 'the danger and disadvantages that would arise if our Society, following too closely in the steps of our parent Institution, were to assume too professional and technical a character.'[24] He continually pressed the matter, and in 1881, to indicate the breadth of its interests, the Society of Telegraph Engineers became the Society of Telegraph Engineers and of Electricians, 'electrician' being the term then used to indicate a man of science rather than an artisan.

It did not work. The full ambitions of the STE's founders, in regard to electrical science, were never realized. The next change of name, in 1888, indicated that the Society of Telegraph Engineers had become reconciled to the idea of taking its place alongside the Institution of Civil Engineers and the Institution of Mechanical Engineers as a professional association which contained a learned society rather than a learned society with professional affiliations.

Acknowledgement

An earlier version of this paper was presented to the Royal Institution's 'Scientific London' meeting in 1982.

Notes and References

1. The first woman member of the Institution, Hertha Ayrton, was admitted in 1899. No other woman was admitted until 1916.
2. *Companion to the Almanac*, 1843, 1.
3. See relevant chapters in Crosbie Smith and M. Norton Wise, *Energy and Empire: A Biographical Study of Lord Kelvin*, (Cambridge, 1989).
4. C.W. Siemens, 'Presidential address', *J. Soc. Telegraph Eng.*, 1873, 1: 19–31.
5. For a discussion of this point see James A. Secord, 'Extraordinary experiment: electricity and the creation of life in Victorian England', in David Gooding *et al.* (eds), *The Uses of Experiment: Studies in the Natural Sciences*, (Cambridge, 1989), 337–83.
6. Quoted by G.F. Fitzgerald, 'Inaugural address by the Chairman', *J. Inst. Elec. Eng.*, 1900, 29: 394–405.
7. Charles Webber, 'Note', *J. Soc. Telegraph Eng.*, 1893, 21: 543.
8. Charles Webber, 'Obituary address of Sir Francis Bolton', *J. Soc. Telegraph Eng.*, 1888, 16: 3.
9. C.W. Siemens, op. cit. (4), 27.
10. Ibid., 22.
11. Ibid., 29.
12. Society of Telegraph Engineers, 'Rules and Regulations', *J. Soc. Telegraph Eng.*, 1873, 1: 10–16.
13. C.W. Siemens, op. cit. (4), 29.
14. Latimer Clark, 'Remarks', *J. Soc. Telegraph Eng.*, 1873, 1: 32–3.
15. Cromwell Varley, 'Remarks', *J. Soc. Telegraph Eng.*, 1873, 1: 34.
16. Curiously, William Thomson's membership in the STE is not discussed in *Energy and Empire.*
17. C.W. Siemens, op. cit. (4), 21–2.
18. Latimer Clark, 'Inaugural address', *J. Soc. Telegraph Eng.*, 1875, 4: 21.
19. H. Schutz Wilson, 'Report on the President's conversazione', *J. Soc. Telegraph Eng.*, 1873, 1: 226.
20. Society of Telegraph Engineers, op. cit. (12), 10.
21. Ibid. Here 'concur' is used in the sense 'to combine in action, to cooperate'.
22. Rollo Appleyard, *The History of The Institution of Electrical Engineers*, (London, 1939), 47.
23. C.W. Siemens, op. cit. (4).
24. Society of Telegraph Engineers, 'Council minutes', 1879, 110–11.

An Appraisal of Fleeming Jenkin (1833–1885), Electrical Engineer

C.A. HEMPSTEAD

> Two years have passed since Fleeming was laid to rest beside his father . . . and the thought and look of our friend still haunts us. (Robert Louis Stevenson)[1]

INTRODUCTION

On 12 June 1885 Henry Charles Fleeming Jenkin died: septicaemia, the enemy of surgery in the late nineteenth century, had claimed the life of a 'remarkable man'.[2] Fleeming Jenkin, born in March 1833, in the decade in which Faraday's researches suggested the promise of electrotechnology, took up engineering as his profession and devoted much of his life to the formation, development and teaching of electrotechnology. By the age of 32 he had been elected to the Royal Society and a year later accepted a Chair of Engineering at University College London. In 1868 he was appointed the first Professor of Engineering in the University of Edinburgh, where he remained until his death. After education at the University of Genoa, he served with a variety of engineering houses, until in 1859 when working for Newall's at Birkenhead he came to the notice of William Thomson. From then on his professional and personal life flourished: forming partnerships with various colleagues, he became a successful consulting engineer, mainly in sub-oceanic telegraphy. The difficulties associated with this new technology allowed Jenkin an entry into the 'science of engineering', where in conjunction with men such as James Clerk Maxwell he laid a sound basis for the development of electrotechnology.

Jenkin's interests did not, however, stand solely in the practice and propagation of engineering. He wrote on literature and drama; his essay on the theory of evolution gave Darwin considerable food for thought; he produced several lengthy articles on political economy; an essay on Lucretius was met with approval; concern about living standards led him to compose a book on the subject of *Healthy Houses*; and he acted as an effective, but amateur, producer of plays.

In spite of his talents and his contribution to engineering in Britain in the middle of the nineteenth century Jenkin is, by and large, unrecorded in the history of technology; indeed, as Buchanan has so perceptively remarked for other 'prominent engineers' of the 'post-Heroic Age',[3] Jenkin falls into the 'penumbra area of relative obscurity'.[4] Buchanan has pointed out that while we possess a host of publications concerning the 'Great Men of Engineering',[5] particularly those active before *c.* 1860, of those of the latter half of the century we know very little. Indeed, Buchanan presents us with a significant question when he wonders, 'what was the quality of this anonymous army of engineers?' The intention of this paper is to bring one prominent engineer, at least, out of the 'penumbra area of relative obscurity' into a more brightly illuminated region.

Apart from obituaries, only one study of Fleeming Jenkin's life has been made: a 'Memoir' written by Robert Louis Stevenson, and published with a selection of Jenkin's writings in 1887.[6] Various aspects of his work have been commented on from time to time, but no detailed examination of his life and work has been undertaken. In the short compass allowed here I can merely touch on the complexity of Jenkin's character and work, and will be highly selective in my sources. Mainly, I will be reporting and commenting on previously unconsidered aspects, with a view to establishing: (a) the character of the man; (b) his metaphysical and epistemological position; and (c) his contribution to engineering. I will be omitting his essay on evolution, his work on political economy, his interest in healthy housing, his dramatic and literary writings, and his consulting and business interests (except insofar as they impinge on the points above). These have been given some consideration by other writers, and some indication as to the impact and originality of Jenkin's thought can be gathered from these works.[7]

THE CHARACTER OF THE MAN

The temptation to read the 'hidden messages' in Jenkin's own writings has been resisted, although his letters, in particular, give the impression of a lively if at times abrupt manner.[8] Rather the choice has been made to examine Jenkin through the eyes and minds of a few of his contemporaries. There are so many points of agreement between those who wrote on his personal qualities that we may, fairly certainly, argue that they were presenting a reasonable picture of their subject. In my consideration of Jenkin's nature I shall compare the opinions of Stevenson, J.A. Ewing and various, usually unsigned, obituaries.[9]

Shortly after Jenkin's death moves were made to publish a collection of his writings, to which Stevenson added a biographical Memoir. The compilation, by no means complete, was edited by Sidney Colvin and J.A. Ewing and published in 1887. Stevenson's desire to honour his dead friend with an obituary was expressed to the widow in June 1885; for Stevenson of all men 'never knew a better one nor one more lovable'.[10] Yet the good early relationship might well have been soured by Stevenson's attitude to Jenkin's engineering. Their first meeting had occurred 'by chance', as is delightfully recounted by Sir Alfred (J.A.) Ewing in *An Engineer's Outlook*,[11] but later Stevenson enrolled for a course of lectures given by his friend. Stevenson, however, did not take professors seriously, regarding them as jokes, with 'Fleeming [being] a particularly good joke'.[12] Stevenson 'was not able to follow his lectures' but 'dared not misconduct himself' under the stern eye of Fleeming Jenkin. Even so, Stevenson considered that Jenkin did not have the presence of a professor: 'no one, least of all students, would have been moved to respect him at first sight.'[13] Unable to exercise the solace of misbehaviour Stevenson absented himself from Jenkin's classes.

However, Stevenson required a certificate of attendance. He was confident that he could persuade Jenkin to provide him with one, for he 'was a master in the art of extracting a certificate even at the cannon's mouth,' but Jenkin was not easily persuaded. It went against his morality to express himself satisfied when, in fact, he was not. Nevertheless, Jenkin reluctantly found a 'form of words' but, chastened, Stevenson left 'with [the] certificate in [his] possession, but with no answerable sense of triumph'.[14] There was a seventeen-year gap between teacher and student and this incident might well have left it unbridgeable, but the event was, for Stevenson, 'the bitter beginning of my love for Fleeming; I never thought lightly of him afterwards.'[15] Perhaps excusing his own shortcomings, the younger man implied that his teacher was, in fact, the less mature: energetic and enthusiastic, but remote and unable to understand 'the mingled characters of men'.[16] According to Stevenson, when they first met, the Professor 'had still something of the Puritan, something of the inhuman narrowness of the good youth'.[17]

Stevenson never knew a 'more lovable' man, but rarely was 'lovable' the first impression that Jenkin made. He was disputatious, revelling in argument, 'jubilant in victory, delighted by defeat'.[18] Jenkin was, wrote Stevenson, 'a Greek sophist, but a British schoolboy'.[19] Jenkin would dispute about anything: 'giving and taking manfully . . . with exuberant pleasure; speaking wisely of what he knew, foolishly of what he knew not . . . picking holes in what was said even to the length of captiousness.'[20] Those who 'truly knew and loved him'[21] were fearful of introducing him to their friends. The results of such meetings were often to make his companions wonder why a 'man so lovable thwarted love at every step'.[22]

A combative nature does not make for easy first impressions, but Jenkin softened as he grew older. To his pleasure he discovered that he began to be asked to 'come and dine', an experience that came new to him as he approached fifty years of age.[23] However, while late popularity smoothed his character he remained 'a bit of a porcupine to the last, still shedding darts'.[24]

In 1871 J.A. Ewing began his engineering studies with Fleeming Jenkin, and started on what was to be a distinguished career: Professorship at Cambridge, Director of Naval Education, FRS, the Creation of Room 40, a knighthood.[25] His essay 'The Fleeming Jenkins and Robert Louis Stevenson'[26] described Jenkin as an 'inspiring teacher of engineering', with an 'interest in . . . all that makes up life . . . unbounded'.[27] Ewing recalled the difficulties that Jenkin had faced in his early years, which he had overcome 'with indomitable resolution', and that by the age of forty he 'had won a place in his profession equivalent . . . to that of a bishop or a judge'.[28] When Ewing met him, 'he had force, maturity, had done much and thought much'.[29] Jenkin had a 'grasp of essential principles and [a] flair for turning them to practical account, [and] his aptitude for scientific research, made him a acknowledged pioneer.'[30] Ewing's reminiscences confirm his disputatious nature and also reinforce Stevenson's recognition that Jenkin's methods did not always meet with favour. For Jenkin: 'revelled in friendly disputation: would always take hard knocks with unruffled temper, and counted—not always prudently—on his disputants doing the like.'[31]

Ewing in his short essay presents a very similar picture to that of Stevenson insofar as Jenkin's character was concerned. They differed only in their assessment of his engineering lectures. To Stevenson, they were incomprehensible;[32] to Ewing, Jenkin was an 'inspiring teacher'.[33] But Stevenson was a 'sedulous truant'[34] whereas Ewing was 'a serious—a very serious—student of engineering'.[35] Jenkin, however, was esteemed by them, and in his turn encouraged and supported them in their chosen careers.[36]

Obituaries have been of some assistance, but as they tend to honour rather than assess the dead it would be too much to expect them to give a complete picture. As many were unsigned we cannot be certain of the provenance of the information they convey. They could have drawn on limited sources. However, the views of the obituary writers concur on the range of Jenkin's interests, and the sharpness of his intellect. Typical are the opinions expressed in the *Electrical Review*:

> Professor Jenkin was a man of remarkable power, and even more remarkable variety. As an accomplished linguist, a litterateur, a critic, a metaphysician, a sportsman, no subject seemed to come amiss. Whether the talk was of technical education or the costume of the Greeks, of French plays or the freedom of the will, Prof. Jenkin was sure to contribute a fact, a theory—a paradox perhaps—but still something fresh and suggestive.[37]

Yet the writer saw him as a 'distant' man, making his character known only to a few, for: 'only his intimate friends really knew him, and his intimates were for the most part specialists in literature or art, or in anything rather than the subjects that formed the main business of his life.'[38]

While the lengthy, and apparently original, obituary in the *Proceedings of the Institution of Civil Engineers* is largely concerned with Jenkin's engineering work, some comments are made which illuminate his personality. For example, he was 'the best talker in London'; he knew 'more of the construction of plays than any man in England'; he was 'something of an artist and an actor'; the 'flash of his restless intelligence could be turned to lighten up any subject'. He spoke 'with a dogmatism which was half-jesting, half serious', but the dangers inherent in versatility were 'tempered by strong common-sense' and by his power of intense concentration.[39] The closing paragraph of this obituary summarizes his character as it has been described here:

> Outside of his family he was a man easy to admire, but hard to know: it was only after long acquaintance that one entered into the inner chamber of real intimacy. Many men who thought they knew him never got so far. . . . His transparent honesty would tolerate no shadow of a lie, least of all in himself. . . . Other men have achieved higher eminence and done more enduring work, but to few is it given to leave a more various impress, or to have their memory cherished with a tender regard.[40]

METAPHYSICS AND EPISTEMOLOGY

Jenkin did not write explicitly on these philosophical issues, but his predilictions were expressed implicitly in his scientific and technical writing. However, two essays, 'Lucretius and the atomic theory' (1868)[41] and 'A fragment on truth' (1884),[42] demonstrate more clearly his attitudes to metaphysical and epistemological principles.[43]

Jenkin's engineering activities and his membership of the British Association's Standards committee brought him face to face with problems of knowledge insofar as it related to the recognition of the existence of electrical quantities. It was clear to him that while electrical effects were readily demonstrated, the 'entities' that underlay the appearances were not easily to be defined or comprehended.[44] His interest in electrical theory and his associations with Thomson and Maxwell brought him up against the ontology of fields, and their conflict with corpuscularian theories. He stood at a watershed in the metaphysics of the micro-world, for as Harré reminds us, Faraday's understanding of electromagnetism was forcing a reconsideration of the existence of a continuum. Thus did ontological issues present themselves to him. He was led to the conclusion that there existed several competing descriptions of the world; thus he became interested with the problems of epistemology and truth.

Jenkin's review of Munro's version of Lucretius' writings was concerned not so much with the Roman writer's ideas *per se* but rather with how they allowed him to examine the 'ultimate constituents of matter'. The essay was written with a good knowledge of the history of science, which he saw in terms of the long existence of two competing theories: continuity or discontinuity; atoms or plenum. Presenting a typical 'Whiggish' view of history, he argued that in ancient and medieval times Aristotelian continuity held sway, to be replaced by a corpuscularian theory after Newton. However, from the end of the eighteenth century, following the writings of Boscovich,[45] there was a re-emergence of the ideas of a plenum, ideas that were strengthened by Faradayian field theories. In 1868 he considered the time was ripe to consider: 'what the real tenets of Lucretius were, especially now that men of science are beginning, after a long pause in the inquiry, once more eagerly to attempt some explanation of the ultimate constitution of matter.'[46]

Jenkin was clear that atomic theories were considered antitheistic, but this question held no interest for him. He was more concerned to examine

Lucretius' ideas in the light of modern knowledge, and to demonstrate how reasonable they were. He reviewed Greek views on the ultimate constitution of matter, but concluded: 'Of all the subtle guesses made by the Greeks [*sic*] at this enigma, one only, we think, has been fruitful, . . . that . . . expounded by Lucretius.'[47]

One by one Jenkin rewrote Lucretius' conceptions of the actions of atoms to match descriptions with perceptions, arguing that 'natural phenomena are subject to definite laws', but he pointed out that 'Lucretius fails to perceive that definite physical laws are consistent with the existence of God', for 'the very conception of a law suggests a lawgiver'.[48] Bringing things up to date, Jenkin introduced spectrum analysis and the laws of chemical combination, but saw these as 'simply special applications of [Lucretius'] general theorem; if matter really obeys definite unchangeable laws, the ultimate materials employed to make matter must themselves be definite and unchangeable.'[49]

The persistence of Lucretius' atoms in spite of the waxing and waning of physical objects was likened to the Law of Conservation of Energy, and Jenkin showed how the concepts of *vis viva* and elasticity were implicit in Lucretius' atomism. Lucretius' acceptance of absolute space and his opinion that in the 'swerving' of atoms and their interaction lay the origins of all that we see, was not, in Jenkin's opinion, compatible with Newtonian dynamics. This did not allow absolute space, and did not permit random, uncaused 'swerving'. Jenkin appears to have required a prime mover in *his* universe, for Lucretius' atoms 'were described as in deadly stillness—a death from which no life could spring, a rest from they could never swerve until inspired with power from a source of life.'[50] Of course, uncaused 'swerving' was a weak point in the Lucretian doctrine, and had been recognized as such long before Jenkin. However, he thought that the atomist's 'conception was not stupid, it was simply false, as all physical explanations of the origin of energy and matter must be.'[51] Jenkin saw that 'swerve' could touch upon 'free-will' but considered it impossible 'that natural science will ever lend the least assistance towards answering the Free-will and Necessity question'.[52] Nor could any insight be given into the nature of a First Cause. But considerations of such questions, important as they were, diverged from his main interest.[53]

The ancient opponents of atomic theories were given short shrift. For example, 'Aristotle and his followers got entangled in the "snare of words", to use Hobbes' language, and their teaching led to little or no progress in what we call science.'[54] Reflecting the current views of the history of science, Jenkin thought that it was not until the seventeenth century that serious, scientific objections to the atomic theory arose, in particular in the work of Descartes and Leibniz. Descartes's ideas were ludicrous, for 'his laws of motion are false, and he knew it, but says we must judge from our experience of gross matter; and yet this man insisted on clear conceptions as the very test of truth.'[55]

Jenkin objected to Leibniz's idea of God as 'Court of Appeal', for 'we do not use the argument now in support of circles as more perfect than other figures, and therefore more consistent with Divine wisdom.'[56] But

Leibniz's monads, while interesting and perhaps leading to Boscovich's centres of force, were in 'unfavourable contrast with [the ideas] of Lucretius'. However, the idea that atoms were voids in an ether rather than small, massive particles moving in nothingness was worthy of consideration, particularly as it came from 'the man who claimed to have run a race with Newton in inventing the higher calculus of mathematics, and who enounced the doctrine of *vis viva*.'[57]

Thus in Jenkin's interpretation of the history of science expressed in his essay on Lucretius effective counters to the atomic theory were emerging just as it was being put on a strong empirical and theoretical base.[58] However, in the mid-nineteenth century Jenkin was happy that the two interpretations, continuity and discontinuity, atoms and plenum, were converging, particularly if William Thomson's extensions to Helmholtz's work on vortex rings could be validated. For, 'Having traced the theory of a continuous fluid to its development in the hands of Thomson, we find that this school too has arrived at indestructible elastic atoms as the secondary constituents of gross matter, though they reject the crude atoms of Lucretius as a primary material.'[59]

In the 1860s solid atoms, moving in a void[60] and acting by collision, were untenable as the fundamental and only constituents of the universe. Forces such as gravitation, magnetism and electricity required either the operation of causes acting at a distance across truly empty space, or the assignment of properties to the vacuum. For electricity, magnetism, heat and light Jenkin considered that the existence of an ether was 'almost demonstrated',[61] and that 'Faraday, by proving the influence of the intermediate material in the case of electrical action, by his discovery of magneto-optic rotation, and by showing how lines of force arose in media, rudely shook the theory of attraction and repulsion, action at a distance across a perfect void.'[62] By analogy even gravitation might be explained.[63]

He concluded that there were three distinct atomic theories: these being 'atoms of "solid singleness", . . . atoms due to the motion of a continuous fluid and . . . atoms having the property of exerting force at a distance.'[64] Jenkin's ontological predilictions were for the primacy of atoms and fields, but in 1868 the 'problem of the constitution of matter [was] unsolved'.[65] The nascent kinetic theory of gases and the laws of chemical combination, although suggestive of atoms, were not in themselves conclusive. But as Faraday's observations had provided the evidence for the existence of fields,[66] so it could be expected that it would be possible to validate observationally the true nature of atoms. It would be necessary to provide a scientific explanation, the nature of which was evident in the writings of Thomson, Clausius, Rankine and Clerk Maxwell. Then the 'vast difference between the old hazy speculations and the endeavours of modern science' would be perceived.[67]

On the face of it deliberations on a standard of resistance would not seem to give rise to particularly profound questions of epistemology, but it is clear from the various reports of the British Association committee appointed to decide upon a standard that the nature of electricity demanded a deep consideration of philosophical problems. Harré's excellent introductory study of the philosophy of science indicates the forms of metaphysical ideas tucked away in electromagnetism[68] and discusses the various categories of measuring devices.[69] The recognition, definition and realization of electrical quantities were not as straightforward as those of length, time and weight, being more akin to those of energy and force. That is, electrical quantities were more evidently 'theory-laden'.

While we cannot be certain that the philosophical undercurrents in the British Association publications were those of Jenkin, even though for some of them he was the reporter, it is clear from his 1873 textbook, *Electricity and Magnetism*, that he was well informed on epistemological questions. In its Introduction he wrote:

> Not a single electrical fact can be correctly understood or even explained until a general view of the science has been taken and the terms employed defined. The terms which are employed imply no hypothesis, and yet the very explanation of them builds up what may be called a theory. *The terms cannot be explained by mere definitions, because they refer to phenomena with which the reader is unacquainted* [my italics].[70]

With an obvious circularity, electricity could not be understood without understanding the 'facts', which could not be understood without comprehending the theory in which the facts were embedded. Further, the facts had initially to be stated for they were not immediately evident to the senses, because: 'Many of the assertions cannot be proved to be true, except by complex apparatus, and the action of this complex apparatus cannot be explained until the general theory has been mastered.'[71] In understanding electricity, then, the worker was brought directly up against problems of knowledge, and the eight years of deliberation of the British Association's Standards Committee demonstrate very well the accuracy of Jenkin's contention regarding the establishment of the truth provided by the components involved in the science of electricity.[72]

In the unfinished, and unpublished,[73] 'Fragment on Truth' Jenkin began to record his thoughts on the certainty of knowledge. I have already remarked that some things were more or less definable by the methods of science—the 'truth' of the 'ether', for example—while other questions could not be settled by physics or chemistry—the free will or necessity question, for instance. In 1885, when he began his essay, Jenkin felt that the question could not be evaded, for 'It is admitted on all hands that in all matters, whether of faith, knowledge or perception, we should endeavour to attain truth; to believe truly, know truly, feel truly.'[74] Yet so different were men's conceptions of knowledge that 'a strong desire [had] ever been felt for some criterion or touchstone . . . [to] discern truth from falsehood.'[75] While the desire might be strong it was not to be satisfied, since 'no such touchstone [had] or ever [would] be found, and indeed the desire is rather akin to the vague longing for magic powers than to any healthy appetite.'[76] Absolute truth, thought Jenkin, was, if not non-existent, then certainly unobtainable, except perhaps in the most trivial sense.

Adopting a form of dualism, Jenkin considered that truth was a measure of 'a concordance between some verbal expression and an external fact or between some mental expression and an external fact.'[77] The statements of mathematics provided some kind of truth, but while 'mathematics [could] provide a sure test of success, it was impotent to suggest a theory'.[78] But 'agreement between many minds as to any statement becomes more and more probable as the statement is more and more restricted to the simplest class of facts'.[79] Herein lay 'the supposed superiority of mathematics, and science generally in regard to truths'.[80] However, these truths meant very little, and it was easy to agree on the 'truth of the language by which they are expressed'.[81] Beyond the simplistic descriptions provided by mathematicians and scientists the expression of complex, real situations demonstrated that 'the statements and the facts accord imperfectly'.[82] In practice, the lack of concordance between descriptions, in words or symbols, and the real world of experience would be obvious to 'the mechanic'. He would know the approximate nature of a 'statement and the corresponding fact', but would accept near agreements as useful. For Jenkin the 'facts' were material, or were actions on material things, such as bridges and 'breaking weights', pumps and the power required to operate them.[83]

One might call this form of Jenkin's truth 'realistic dualism'—characterized by a dichotomy between the descriptions of facts and the facts themselves. His alternative definition of truth could be described as 'idealistic dualism'. The facts were of the same type, but the accord now was 'between what seems and what is, between the conception in our minds and the fact which gives rise to that conception'.[84] He could not accept that truth was attainable by examining such concordance. The problem was one of perception, and he could see no means by which a mental impression, a concept to him, could have an identity with the object that allowed the idea to be conceived. Again he turned to a practical argument. The dent produced by a hammer is not the same as the hammer; no two coins are identical, even if struck from similar metal with the same die. Thus if material things, seen as analogies to mental constructs, lacked identity, how much more unlikely was it that perceptions, widely different from one person to another, could represent the actuality of objects? Idealism could not work, except approximately. Jenkin did not object to relative or absolute truth, but concluded that he had 'exposed the folly in any hope of a criterion of absolute truth'.[85]

He did wonder that 'if there were a criterion—is it conceivable that we should ever apply it with exactitude?'[86] He intended to examine at least two contenders: physical measurement and mathematics. Unfortunately, his consideration of the latter was cut off, leaving us with nothing to develop. In his discussion of measurement, while he recognized the existence of a personal equation, from the point of view of his realistic dualism the mind and its sensations could be ignored. For in 'scientific measurement things, not sensations, are compared'.[87] Jenkin did not distinguish among 'things' whose measurements were obtained by 'self-measurers' or 'non-self-measurers'.[88] His 'things' included time, space, matter, energy, and so on; and measurements compared position, velocity, tenacity, and so forth.[89]

Jenkin shared Thomson's opinions regarding the association between knowledge and measurement, and, like him, seemed to be convinced that regardless of 'absolute truth' the numerical relationships between objects could be defined and measured scientifically. He thought that: 'In these regions dispute is settled by an appeal to things not to minds; no criterion of truth is missed, and prolonged difference of opinion concerning the relations measured is impossible, except as regards a small and constantly lessening fringe or margin of the whole part.'[90]

Jenkin's epistemological musings were echoes of much earlier ideas. As we have suggested, his work with the British Association's committee on electrical standards brought the problem of electrical knowledge to the forefront of his thoughts. In the opening section of the paper on Electrical measurements[91] (first written in 1863, and reprinted in 1865, with Maxwell as co-author), a philosophical stance was laid out. Simply it was the aim of the paper to lay out a means whereby 'laws remembered in their abstract form [could] be applied to estimate the forces required to effect any given practical result'.[92] It was important from the point of view of a practising telegraph engineer to be sure in advance of the limits of his devices and techniques. Exact knowledge of electrical quantities was required, and:

> All exact knowledge is founded on the comparison of one quantity with another. In many experimental researches conducted by single individuals, the absolute values of those quantities are of no importance; but whenever many persons are to act together, it is necessary that they should have a common understanding of the measures to be employed.[93]

Jenkin considered what 'things' should be the basis of any standard, and concluded that they should consist solely of what he called 'primary units', that is, those of mass, length and time,[94] because 'a system in which every unit is derived from the primary units . . . is the best whenever it can be introduced . . . it bears the stamp of the authority, not of this or that legislator or man of science, but of nature.'[95] While in this publication the manner in which units could be realized was not discussed, the practical problems were not ignored by the committee. Two types of problem were clearly identified: first, the definition and determination of the standards themselves, and second, the specification and manufacture of the sub-standards. Jenkin's understanding of the significance of these difficulties is made explicit in his reply to Werner Siemens on the subject of the definition and determination of resistance.[96] In this work Jenkin administered a firm rebuke to Siemens, who apparently did not grasp the requirements of either a standard or a sub-standard.

Jenkin's metaphysics recognized the need for atoms and a plenum, and he saw the resolution of the dichotomy in terms of the ether, which could 'condense' in vortex rings to form centres of force. Thus could the world be explained. However, his epistemology was empirically based: exact, perhaps true knowledge derived from measurements associated with existent 'things'—'things' that were not hypothetical but real. Reality was assured by the physical existence of instruments and artefacts. While he held this as a general principle, it was in particular true in his electrical engineering. Nevertheless, as he wrote in the Introduction to his *Electricity and Magnetism*, while the instruments themselves become outdated the principles 'on which the construction and use of these depends is permanent, depending on no hypothesis'.[97]

THE CONTRIBUTION TO ENGINEERING

Jenkin's professional interests were wide. His published work, dealing with bridge construction, mechanical engineering and sanitary engineering, as well as electrical engineering, demonstrates his interests, and his patents show his concern with engineering as a source of income. His contribution to engineering did not end with its practice, for as Professor of Engineering at Edinburgh he took his teaching duties very seriously, encouraging both undergraduates and 'research' students[98] in their learning—for Jenkin the post of Professor was not simply a useful title, or a sinecure.[99] While the bulk of his work was concerned with telegraphy in general and submarine telegraphy in particular, from early in his career he took great interest in technical innovations and 'new' scientific ideas that might provide practical applications.[100]

Jenkin was fortunate in his talents, education, career choice and associates. Intellectually he was broadly gifted; and, according to his father, his gifts were manifested early. In an unpublished manuscript[101] composed in June 1865 Charles Jenkin wrote that, in the Edinburgh Academy, 'he early distinguished himself amongst his Class Fellows especially in Mathematics and Classics. He bore away many Prizes.'[102] Force of circumstances took the Jenkin family via Frankfurt and revolutionary Paris in 1847–8 to Genoa, where Fleeming enrolled in the University. Here he displayed strength of purpose, not least evident in his mastering Italian.[103] His father 'observed how closely [he] applied himself to all the Instruction there, Mathematics, Classics, Rudiments of Engineering, Physics', and noted that 'Fleeming [had] always shown a talent for Drawing. Here he attained great perfection and received [a] Medal and his Degree of Batchelor [*sic*] of Arts.'[104]

His route to submarine telegraphy was directed by a series of attachments to different engineering concerns. Beginning in Genoa, where he worked for a Marseilles-based firm, and passing through Fairbairn's works in Manchester, he found himself in the latter half of the 1850s working at Newall's of Birkenhead. In the course of this engineering 'journey' Jenkin learnt the practice of mechanical engineering, the importance of surveying and some railway engineering—his theoretical and practical training had been broad.

With Newall's he settled into submarine telegraphy and in 1859 his duties were concerned with the Red Sea cable.[105] It would appear that he was responsible for a range of activities concerned with stowing and testing the cable, and his father remarked upon this work:

> As this [submarine telegraphy] was new to Engineering I should think much ingenuity must have been required to set up the necessary machinery, afterwards the stowing it aboard a Vessel, The Paying out and was (I believe) entirely left to Fleeming. I am not qualified to judge what amount of Intelligence was manifested in manufacturing so extensive a submarine Cable, The Testing Instruments of its powers of Conductivity through more than 2000 miles. From this time (I believe), Professor Thompson [*sic*] of Glasgow University made acquaintance, and they continued closely in their special science.

Charles Jenkin proudly ended his biographical note thus: 'In addition to many Honours, He has this year June 1865 been chosen a Fellow of the Royal Society.'

While Jenkin's father felt unable to comment on his son's abilities his employers and colleagues were competent so to do. In 1859 Lewis Gordon, of Liddell & Gordon, visited William Thomson to discuss some instrumental aspects of submarine telegraphy. Stemming from what he had seen, Gordon remarked that he 'would like to show this to a young man of remarkable ability'; the young man was Fleeming Jenkin. Jenkin travelled to Glasgow, and impressed William Thomson with his abilities. Thomson was 'much struck, not only with his brightness and ability, but with his resolution to understand everything spoken of, to see if possible thoroughly through every difficult question, and (no *if* about this!) to slur over nothing'.[106] Thomson's and Jenkin's association lasted until the latter's death. In 1859 Jenkin formed a partnership with H.C. Forde, which was becoming financially successful by the time he took up his Professorship in Edinburgh. With this appointment the association was dissolved, but by then Jenkin had formed an agreement with Thomson and Cromwell Varley.[107]

There is no evidence that he planned his career and his partnerships, but if he had he could hardly have done better. His education at Genoa introduced him to the theoretical side of engineering, and gave him a lifelong belief in the power of mathematical argument. His series of 'apprenticeships' provided him with a broad complement of engineering and management skills that he was to use to good effect in his submarine cable-laying 'adventures'. His association with Thomson, Forde and Varley put him into intimate contact with the development of a new technology—that of submarine telegraphy. His contribution to this mid-nineteenth century 'high technology' was highly valued—not least by William Thomson.

The successful introduction of new technologies not only requires the right economic and social conditions but needs the presence of talented workers.[108] As I have argued elsewhere,[109] established technologies can be maintained by 'rule books', of course used sensibly and with knowledge; but new technologies have no such guides. They have to be written. And the means by which they are composed rest upon the identification of problems, the definition and realization of parameters that enable the problems to be solved, and plenty of practice—supervised by the authors of the rule books.[110] Once the practice has been established on a sound footing, then the authors can, and often do, withdraw from a close association with the technology.[111] It is not infrequently the case that while new technologies involve the use of pre-existing techniques they push the older ideas to the limits of their applicability.[112] Thus it was with submarine telegraphy.

It is not my intention here to rehearse the history of submarine telegraphy, for many works have been composed on the subject. However, it is well worth remarking on the broad aspects of its story. In 1851 Dover and Calais were connected by a cable, which, eventually, operated successfully for a number of years. In the next few years relatively short lengths of cable were stretched across narrow seas—the North Sea, the Irish Sea, and so on. Apparently no severe problems were met with—although intelligible signals were not always easy to transmit—encouraging financiers and engineers to set in hand the project of designing, manufacturing and laying a transatlantic cable. In 1857 and 1858 the connection was attempted. As is well known initial successes were short-lived, for the cable failed. Yet the promise of long-distance signalling had been demonstrated. However, the failure of the Red Sea cable was altogether more important for the British Government.[113] The immediate outcome of these signal failures was a Privy Council Committee of Enquiry, whose purpose was to identify the causes of failure and to ensure that, in future, cable-laying would be successful.[114]

In another paper[115] I have suggested that the Committee acted as a legitimating body, underlining and supporting the opinions of those engineers and scientists who already *knew* what the problems were, and *knew*, in principle at least, what needed to be done to solve them. The pages of the Report of the Committee detail the manifold stances of the different engineers, but a small group of men, including Jenkin, made their position abundantly clear, arguing for the alliance of theory, careful laboratory measurements, and well controlled, large-scale practice.[116]

From 1861 the proposals of the Committee, embodying the ideas of this limited group of engineers,[117] were used in submarine telegraphy. Cables of increasing length were constructed, and increasing reliability was achieved. Techniques were developed for recovery of damaged cables, and increasingly sensitive detecting instruments were designed. By 1865 a new attempt was made on the Atlantic, and in 1866 a successful connection was completed.[118] By 1870 the 'rule book' for submarine telegraphy had been written; there were to be no substantial changes for two generations.[119]

The submarine telegraph system made its contributions to international trade[120] and assisted British Imperialism,[121] but most importantly, from the point of view of this paper, it allowed the definition and realization of a system of electrical units. The investigations needed to establish this system helped develop a variety of instruments, and *en passant* encouraged a new science. The study of the electrical properties of solids and attempts to comprehend transmission of signals along cables helped to validate

Maxwell's theory of the electromagnetic field.[122]

Three electrical parameters were of profound significance in the theoretical and practical understanding of insulated submarine cables: the insulating qualities of gutta-percha, the resistance of the copper conductor and the capacity of the cable. Thomson was in no doubt that Fleeming Jenkin made if not the first, then the most important, contributions to the definition, measurement and understanding of these quantities in this formative period. In his 'Note on the contributions of Fleeming Jenkin to electrical and engineering science' (1887)[123] Thomson made the following observations: '[In 1859 Jenkin] began definite scientific investigation of the copper resistance of the conductor, and the insulating resistance and specific inductive capacity of its gutta-percha coating . . . and he was the very first to introduce systematically into practice the grand system of absolute measurement founded in Germany by Gauss and Weber.' Thomson was in no doubt when he noted that: 'The immense value of this step, if only in respect to the electric telegraph, is amply appreciated by all who remember or who have read something of the history of submarine telegraphy.'[124]

In his lifetime Jenkin, as author or co-author, published some forty papers,[125] and applied for or was granted thirty-five patents. The subjects of these serve to underline the width of his interests. The majority of his papers, thirty, and about half of the patents, eighteen, dealt with electrical engineering or related topics.[126] While there is no particular pattern to his publications, with two exceptions between 1859 and 1869 and with three exceptions from 1878 until his death he wrote exclusively on electrical matters.

His papers on electrical topics exhibited considerable originality of thought and method, and were expressive of a first-class mind; at the same time they demonstrated his excellence as an engineer. However, they dealt with matters well within the 'mainstream of innovation'.[127] For example, his writings on submarine telegraphy concerned themselves fairly and squarely with the problems of that emergent technology; but the problems that he addressed were often suggested by others, or by evident force of circumstance. For example, his work on the speed at which meaningful signals could be transmitted along insulated cables[128] stemmed directly from working practice, the theories of Thomson and disputes entered into by Wildman Whitehouse.[129] Similarly, although it was acknowledged that Jenkin's efforts on behalf of the British Association's Standards Committee were second to none, the nature of the work and the method of its reporting do not allow an independent assessment to be made of an individual's contribution. Indeed, this point was recognized by William Thomson, who saw that as far as the realization of electrical standards was concerned, 'it can scarcely be known generally how much is due to Jenkin'.[130] My claim here is that in his publications on submarine telegraphy and electrical standards he was, with skill and originality, working along clearly defined lines, albeit laid in new ground. And so it was with his later publications.

Writing on the phonograph, electrical lighting and his system of 'telpherage' he was following paths indicated by others. He was aware, for instance, of the scientific value of the phonograph, as well as its commercial significance;[131] no electrical engineer of the 1870s and 1880s could escape the problem of the 'subdivision of the electric light';[132] and many had considered the possibilities inherent in electric traction and haulage.[133] In these areas Jenkin's contributions were reckoned to be significant by his contemporaries. Furthermore, it was directly as a result of his involvement with submarine telegraphy that he served on the British Association's Standards Committee from 1861; his value to electrical engineering was recognized by his appointment as a juror on electrical instruments at the London Exhibition of 1862[134] and at the Health Exhibition in London in 1884; and his contribution to engineering in general was recognized by his appointment as a juror in engineering at the Paris Exhibition of 1878.[135] Of course, his election as a Fellow of the Royal Society in 1865 at the age of 32, his appointment as Professor of Engineering at University College, London in 1866, and his acceptance of the Professorship of Engineering at Edinburgh University in 1868 are further testimony to his worth.

All these things are significant for our appraisal of Jenkin as an engineer, but there were many like him who made worthy contributions to their chosen professions and attained similar prominence in their own time. Indeed, if we are to approach an answer to Buchanan's query,[136] these men must also be studied, as must Jenkin's mainstream work, in order to arrive at unequivocal assessments of their contributions and importance. In the case of Jenkin it is, however, possible to measure him not only by the major part of his writings, but also by the exceptional pieces. By and large, these pieces were published between 1869 and 1877. They dealt with structures and machines—and they marked significant qualities.

In 1864 Maxwell published his 'theory of reciprocal figures' and demonstrated that the method could be applied to determinate structures using examples that left 'nothing to be desired by the mathematician'.[137] Jenkin was concerned to bring it to the attention of engineers that the method could be applied in 'real' situations, and in a paper read to the Royal Society of Edinburgh on 15 March 1869 he presented his audience with the theory and showed how it could be applied.[138] However, more often than not frameworks were indeterminate, and different methods were required.

Jenkin had first come across the problems of indeterminate structures in 1861 when he was pondering the 'correct form' of an iron arch. He first touched on the subject of bridges in a postscript to a letter (22 February 1861) to William Thomson.[139] The postscript contains a sketch of an arch bridge constructed from hinged members arranged so that all the forces should be compressive. He was very pleased with this idea, patented it in 1861, and concluded that: 'Nothing could be easier than to calculate geometrically the strain on every component part of the bridge, however loaded—turn it upside down and it makes the true form of suspension bridge—as stiff as you please.'[140] However, his youthful enthusiasm—he was 27—was dampened a little when he discovered he could not, in fact, calculate the stresses and strains. By 1861 Jenkin, now an associate of

Maxwell, raised the subject with him, with the result that Maxwell 'discovered and published in the "Phil. Mag.," May 1864, a method by which the resultant and all stresses in framed structures could be positively determined'.[141] The method was that of virtual velocities.

Jenkin published his mature thoughts on bridges first in the *Transactions of the Royal Scottish Society of Arts* in 1870, where he pointed out that girders, arches and suspension bridges were in fact members of the same class. Girders and arches differed one from the other only in the observer's perception, while suspension bridges, as he had clearly stated in his letter of 1861, were merely reflections of arches. A satisfactory theory was wanted: in Jenkin's opinion, he had found the required methods in Maxwell's papers: the theory of reciprocal frameworks provided useful graphical techniques; the use of virtual velocities allowed *all* frameworks to be subject to calculation. On the graphical methods of treating stresses the *Minutes of the Institution of Civil Engineers* put forward the opinion that Jenkin was instrumental in introducing the graphical method to engineers and that 'by making it a feature of his teaching, . . . did much to acquaint the profession with its use.'[142]

Jenkin was attracted to graphical methods, apparently having a dislike for algebra, and applied them to his work on political economy as well as to his brilliant study of machines, published in 1877.[143] It is a matter of regret that in this paper I cannot carry out a detailed textual and contextual analysis of this work, for the published paper is some sixty pages in length. However, it was seen to be of great worth and was given immediate recognition by the Royal Society of Edinburgh, being awarded 'its highest distinction—the Keith gold medal'. The most striking feature of the paper is its generality, a generality that was based on a careful and thoroughgoing analysis. Jenkin stated clearly his method: 'we may represent any machine, at any given instant, by a frame of links, the stresses in which are identical with the pressures at the joints of the machine. . . . This . . . frame may be so drawn as to represent the machine either rigorously, . . . or approximately, omitting some conditions under which the machine works.'[144] Continuing with the generality of the method, he wrote: 'for all machines (in which the motion can be represented as in one plane) the dynamic frame is of one type, either simple or compounded'.[145] Jenkin claimed that his 'dynamic analysis of machinery' was novel, although consistent with other analyses.[146] He then developed his ideas at length, in a manner reminiscent of an axiomatic system and heavily redolent of Maxwell's reciprocal frameworks, and showed how the design of machines could be effected utilizing his graphical techniques.[147]

The quality of Jenkin's contribution to the theory of machines was recorded in many obituaries and appreciations. The *Proceedings of the Royal Society of Edinburgh* thought that his paper was 'a continuation, full, however, of originality, of the subject treated in Reuleaux's *Kinematics of Mechanism*'.[148] The *Proceedings of the Institution of Electrical Engineers* considered that he had laid 'the foundation of an entirely new dynamical theory of machines, which is as much in advance of the merely kinematic analysis of Reuleaux as that was in advance of the fragmentary treatment which preceded it'.[149]

William Thomson was less enthusiastic, but thought that Jenkin's approach was 'a continuation of the subject treated in "Reuleaux Mechanism", and supplied the elements required to constitute ... a full machine receiving energy and doing work'.[150] The *Minutes of the Proceedings of the Institution of Civil Engineers*, however, thought that of all Jenkin's papers, that on Graphic Methods was the most remarkable. The ideas were based on Reuleaux's ideas 'but the turn taken by Jenkin's thoughts was so original that, did he not acknowledge it, the influence of Reuleaux would scarcely be suspected'.[151]

In 1866 Jenkin was appointed to the Chair of Engineering at University College London. Nothing can be said of his time there, for nothing appears to have been kept; even Stevenson did not refer to this period.[152] However, he presumably made an impression for in 1868 he was elected to the Chair of Engineering at Edinburgh, where he was the first holder of that post.[153] R.M. Birse points out that of the candidates for the post two were outstanding, Rankine and Jenkin, but gives as his opinion that the terms of the chair, with an emphasis on industrial applications, favoured Jenkin. Buchanan considers Jenkin's appointment to have been significant, for Jenkin followed the 'Rankine' system in which theoretical and practical instruction were to be integrated into engineering education. Buchanan argues that Jenkin came down strongly in favour of Britain's 'traditional pupilage/practical structure',[154] but points out that Jenkin identified as a major deficiency 'the want of a good knowledge of the theories [as they] affected ... practice'.[155] Jenkin, as we have briefly noted, did not lack such knowledge and structured the curriculum and the lecture courses at Edinburgh to reflect his own experiences and beliefs.

In two papers separated by fifteen years he outlined his approach to technical education. In 1869, drawing a clear distinction between 'the scientific education of artisans and foremen, on the one hand, and manufacturers and engineers, on the other hand',[156] he pointed out that the former required courses of instruction that would enable them to read and interpret drawings, while the latter over and beyond their technical knowledge should be the receivers of a 'liberal education'.[157] He recognized, however, that as it stood the British (strictly speaking, the English) education system was in need of much restructuring before technical education would achieve a satisfactory state.[158] In 1884, six years before Edinburgh acquired an engineering laboratory, Jenkin speculated on the role of laboratories in science teaching.[159] He identified two forms of practical instruction: first, that which involved the teacher and assistant in research, and second, that which was essentially the teaching of measurement techniques.[160] From the former the assistants would gather an idea of the form of advanced work, would develop an enthusiasm for the subject and in turn would become leaders. From the latter, each individual (the majority, in fact) would emerge 'a successful manufacturer, contractor, engineer, or farmer'.[161] The majority of students would not acquire 'the hunger for knowledge, the acute logical discrimination, nor the imaginative faculty required for research' which would be the marks of the minority, but could be given an appreciation of the vital importance of quantitative measurements.[162] All students, potential leaders or rank and file,[163] would thus acquire some indication of the meaning of scientific ideas and methods and some knowledge that would prove useful to them in 'after life'.[164] For Jenkin, it was not desirable to separate theory from practice, or the 'high flyer' from the 'rank and file'.

According to Buchanan it was Jenkin and a 'handful of like-minded colleagues' who laid 'the foundations for the transformation, by the end of the century, of the attitudes of the engineering profession towards theoretical instruction and to place beyond dispute the scientific quality of engineering.'[165] For this alone Jenkin can be ranked alongside men such as Rankine, who attempted to modernize and restructure technical education in Britain. However, even if Buchanan is correct, in terms of numbers of engineering graduates from Edinburgh Jenkin's efforts did not meet with marked success. According to Birse, from 1873 to 1884 some twenty students graduated from Edinburgh with a BSc in Engineering, with no more than three emerging in any one year.[166] However, there may have been students other than Robert Louis Stevenson who enrolled on engineering courses with little intention of graduating as Bachelors of Science.

Just as I have used the 'exceptions' to measure Jenkin's contribution to the practice of engineering, so I can look to a few of the more successful alumni of his classes for a measure of his legacy as a teacher. It has long been recognized that a measure of a good teacher, or a good researcher, is the quality of the students attracted to him,[167] and while Jenkin may not have taught many students a few of them formed part of Buchanan's 'handful of like minded colleagues'. Buchanan named, among others, C.F. Jenkin and Alfred Ewing. C.F. Jenkin, the son of Fleeming, became the first Professor of Engineering at Oxford University in 1908 and made a considerable impact on that ancient foundation. Ewing, 'a brilliant and articulate Scots engineer', was Professor of Mechanism and Applied Mechanics at Cambridge from 1890 to 1903.[168] He went on to achieve great distinction, fulfilling the promise that Jenkin saw in him and that was recognized in his selection for professorships in Tokyo, and then at Dundee. In his address to the Class of Engineering at Edinburgh in 1883 Jenkin spoke of these men in proud tones. His son had greatly helped with telpherage, and he informed his audience: 'the first medallist of this class, Mr. R.H. Smith, is Professor of Engineering in Joseph Mason's College in Birmingham, and . . . my friend, Mr. James Alfred Ewing, who was, I think, our youngest medallist, is now Professor of Engineering at Dundee'.[169]

Jenkin went on to refer to 'numerous other successful men' from previous classes, but deprecated his own part in their success. However, he did believe that 'the endowment of the chair [of engineering] . . . had been followed by good results'.[170] While, no doubt, the quality of the men themselves was an important part in their success, the influence of Jenkin on them seems to have been profound. At least that was the opinion of Ewing and some obituary writers—none of whom shared Stevenson's opinion. Almost as an aside it is worth remarking that as part of his contribution to education Jenkin composed his textbook on electricity and magnetism. This work was well received and widely read. Its English-language edition, for example, was for a few years required reading for students following courses in physics at Cornell,[171] and it was translated into several European languages. In his textbook Jenkin not only produced an innovative approach to the teaching of electricity, but also demonstrated how theory and practice could be conjoined—an idea that had long occupied his mind.[172]

CONCLUSION

Having examined Jenkin's character, having outlined his metaphysics and epistemology and having assessed his contributions to the practice and teaching of engineering, I shall now try to give something of an answer to Buchanan's question concerning the quality of one member of that anonymous band of more or less prominent nineteenth-century engineers. In the case of Jenkin I have outlined part of the history of one of those who from the mid-1850s founded and developed the science of electro-technology and the profession of electrical engineering.

It has been argued that Jenkin was well placed to make a significant contribution to electrical engineering in both its practice and promulgation. He was broadly and well educated, with the curriculum at Genoa well fitted to his later life; he was fortunate in his talents, skilful in the arts, in languages and in mathematics; his engineering 'apprenticeship' was varied and detailed, giving him knowledge of a wide range of practices; and he was fortunate to be working at Birkenhead when the technology of submarine telegraphy was in its early stages. The width and depth of his qualities fitted neatly those that are required for the introduction of novel technologies, and his quality was sufficiently obvious to allow him to work with William Thomson. Association with such a man led him to professional heights culminating in his election to the Edinburgh Professorial Chair of Engineering, from which position his powers of communication were evidenced. Throughout his life Jenkin displayed a good grasp of mathematics coupled with an appreciation of the complexities of 'real life',[173] and from time to time he expressed these powers in publications away from his main interests. Thus he published careful studies in political economy, the theory of evolution, the theory of structures and mechanical engineering. In all these his contributions were acknowledged to be profound.

He held a consistent metaphysical and epistemological position. In his work for the British Association and in his discussions with Werner Siemens on the matter of the realization and the determination of a resistance standard his realistic, empirical philosophy is clear; and in his work on Lucretius his ontology is displayed. Near the end of his life he began a study on 'truth' in which his ideas on epistemology were rendered more or less explicit.

Jenkin was at once flexible and persistent. Open to new ideas and anxious to exploit them, he nevertheless followed up some thoughts through a number of years. Thus, from Jenkin's 1861 belief that he had produced the perfect bridge Maxwell developed his ideas of reciprocal frames. Jenkin employed these ideas in 1869 in his final design methodology for girders and suspension bridges, and in 1877 won the Royal Society of Edinburgh's Keith medal for his paper on graphical methods in the theory of machines. Yet he seems to have been somewhat self-deprecating, and perhaps somewhat overshadowed in his own mind by the merits of William Thomson. In his partnership with Thomson and Cromwell Varley he took very much the junior part, and took off his colleagues' hands purely business matters.[174] Whether such an occupation diverted him from technical matters must remain the subject for another paper. It is interesting to note that from about 1870 to about 1877, when it would appear that the partnership was truly flourishing, Jenkin published very little.

His science and engineering experience would seem to have formed his metaphysics,[175] and he was anxious to discuss his ideas with his associates,[176] from whom he seems to have received some encouragement. His careful approach and his epistemology informed his attitudes to education, attitudes which he expressed in lectures, papers and his textbook on electricity and magnetism. In the latter he embodied his ideas on the need for a collocation of theory and practice, although he did not live to see a teaching laboratory established in Edinburgh. He was an inspiring teacher, and bequeathed to the engineering world a number of highly talented graduates, at least one of whom took away with him an appreciation of Jenkin's approach and methods.

Jenkin, then, was a man of high qualities covering a number of diverse activities. In this paper I have only touched on those things that affected directly his engineering, or sprang from it, but it can be concluded that Jenkin's contribution to engineering was sufficiently significant to warrant this study. The full account of Jenkin and his work still remains to be given, but it would appear that his influence spread far and wide from Edinburgh, reaching out, like his submarine telegraph lines, from Britain to the Far East.

Notes and References

1. Robert Louis Stevenson, 'Memoir', in Sidney Colvin and J.A. Ewing, F.R.S. (eds), *Papers Literary, Scientific, etc. by the late Fleeming Jenkin, F.R.S. LL.D.*, (London, 1887), cliv.
2. A.D. Brownlie and M.F. Lloyd Prichard, 'Professor Fleeming Jenkin, 1833–1885, pioneer in engineering and political economy', *Oxford Economic Papers*, 1963, 15: 204.
3. R.A. Buchanan, *The Engineers: a History of the Engineering Profession in Britain, 1750–1914*, (London, 1989), 20.
4. Ibid.
5. Such as Telford, Brunel, the Stephensons, etc. There are a few biographies of more recent engineers, as is indicated in the bibliography to Buchanan's book.
6. Stevenson, op. cit. (1), xi–cliv.
7. D.R. Oldroyd, *Darwinian Impacts*, (Milton Keynes, 1980), 135–7, and James R. Moore, *The Post-Darwinian Controversies*, (Cambridge, 1979), 128–31,

give brief analyses of Jenkin's work and significance. Jenkin's essay is reprinted in D.L. Hull, *Darwin and His Critics: The Reception of Darwin's Theory of Evolution by the Scientific Community*, (Cambridge, Mass., 1974), 303–5.

On political economy see Brownlie and Prichard, op. cit. (2), and J.A. Schumpeter, *History of Economic Analysis*, (London, 1954). In keeping with the 'penumbra' concept, the latest editions of this work have excised references to Jenkin's ideas!

There have been no studies of Jenkin's literary work or his work on healthy houses since Colvin and Ewing, op. cit. (1), except an unpublished final-year dissertation by one of my students at Teesside Polytechnic: T. Nunn, *A Study of Henry Charles Fleeming Jenkin—a Victorian Engineer (1883-1885)*, (Teesside Polytechnic, 1990).

For brief comments on Jenkin's attitudes to consultancy and business, see the monumental work of Crosbie Smith and Norton Wise, *Energy and Empire*, (Cambridge, 1989).

8. Smith and Wise, op. cit. (7), 701.

9. Many of these seemed to be derivative, but two have proved particularly useful, and can be found in *The Telegraphic Journal and Electrical Review*, 1885, 16: 555, and *Minutes of the Proceedings of the Institution of Civil Engineers*, 1885, 82: 365–77. The latter is particularly full.

10. Unpublished and uncatalogued letter in the possession of Lord Jenkin of Roding (formerly The Rt Hon. Patrick Jenkin, MP). The letter is undated, but was evidently written shortly after Jenkin's death. It carries a Bournemouth address.

11. Sir Alfred Ewing, *An Engineer's Outlook*, (London, 1933).

12. Stevenson, op. cit. (1), cxxx.

13. Ibid.

14. Ibid., cxxxiii.

15. Ibid.

16. Ibid., cxxxiv.

17. Ibid.

18. Ibid., cxl.

19. Ibid.

20. Ibid.

21. Ibid., cxli.

22. Ibid.

23. Ibid.

24. Ibid.

25. In various places in *The Engineers*, Buchanan, op. cit. (3) touches on Ewing's career. For details of Ewing's life see the Preface to Ewing, op. cit. (11), vii–xxiv, and A.W. Ewing, *The Man of Room 40: The Life of Sir Alfred Ewing*, (London, 1939).

26. Ewing, op. cit. (11), 248–76.

27. Ibid., 248.

28. Ibid., 250.

29. Ibid., 251.

30. Ibid.

31. Ibid., 251–2.

32. Stevenson, op. cit. (1), cxxx.

33. Ewing, op. cit. (11), xii.

34. Ibid., 248.

35. Ibid., xii.

36. In the case of Stevenson see, for example, Stevenson, op. cit. (1), cxxiv,

for a brief discussion of Jenkin's opinion of his work: in the case of Ewing, see Ewing, op. cit. (11), xii–xiii.

37. *The Electrical Review*, op. cit. (9), 555.

38. Ibid.

39. *Proceedings of the Institution of Civil Engineers*, op. cit. (9), 369.

40. Ibid., 376–7.

41. First published in *North British Review*, 1868; reprinted in Colvin and Ewing, op. cit. (1), I: 177–214.

42. Colvin and Ewing, op. cit. (1), I: 264–8.

43. Colvin referred to Jenkin's non-technological writings as the product of his 'by-labours', ibid., clxxiv.

44. See, for example, the introduction to his textbook *Electricity and Magnetism*, (London, 1873), iii–ix.

45. He expressed his interest in the writings of Boscovich in a letter to William Thomson dated 1867 and held in the Kelvin Collection, University of Glasgow Library, J69.

46. Colvin and Ewing, op. cit. (1), 177.

47. Ibid., 178.

48. Ibid., 179.

49. Ibid., 183.

50. Ibid., 192.

51. Ibid.

52. Ibid., 193.

53. See discussion in ibid., 193–7.

54. Ibid., 198.

55. Ibid., 199.

56. Ibid., 200.

57. Ibid., 201–2.

58. As I shall remark later, Jenkin was thinking of the kinetic theory of gases, and the laws of chemical combination.

59. Colvin and Ewing, op. cit. (1), 206.

60. The 'void', Jenkin was careful to emphasize, was not only devoid of material, but without properties or immaterial entities.

61. Colvin and Ewing, op. cit. (1), 208.

62. Ibid.

63. See discussion in ibid., 207–10.

64. Ibid., 210.

65. Ibid., 211.

66. See also the discussion in J. Clerk Maxwell and Fleeming Jenkin, 'On the elementary relations between electrical measurements', *Philosophical Magazine*, 1865, 29: 440. This paper was a reprint from the *Reports of the British Association*, 1863.

67. Colvin and Ewing, op. cit. (1), 213–14.

68. R. Harré, *The Philosophies of Science*, 2nd edn, (Oxford, 1985), 133, 139.

69. Ibid., 166.

70. Jenkin, op. cit. (44), vi.

71. Ibid., vii.

72. See, for example, the various and lengthy discussions in Maxwell and Jenkin, op. cit. (66).

73. Unpublished, that is, in Jenkin's lifetime.

74. Colvin and Ewing, op. cit. (42), 264.

75. Ibid.

76. Ibid.

77. Ibid.
78. Ibid., 213.
79. Ibid., 264.
80. Ibid.
81. Ibid., 265.
82. Ibid.
83. Ibid.
84. Ibid.
85. Ibid., 266.
86. Ibid.
87. Ibid., 267.
88. See Harré, op. cit. (69).
89. Ibid., for an outline discussion of these things.
90. Colvin and Ewing, op. cit. (1), 267.
91. Maxwell and Jenkin, op. cit. (66).
92. Ibid., 436.
93. Ibid., 437.
94. Ibid.
95. Ibid.
96. H.C.F. Jenkin, 'Reply to Dr. Werner Siemens's paper "On the question of the unit of electrical resistance" ', *Philosophical Magazine*, 1866, 4th ser., 32: 161–77.
97. Jenkin, op. cit. (44), viii.
98. The term 'research student' is somewhat anachronistic, but it serves to define the role of men such as Ewing, who acted as Jenkin's assistants in his research and development work.
99. That it was no sinecure is evidenced by the course Jenkin gave, and that it carried only a relatively small salary. See, for example, R.M. Birse, *Engineering at Edinburgh University*, (Edinburgh, 1983), 100–2.
100. For example, in a letter to William Thomson dated 21 February 1861, Jenkin outlined a new idea for the construction of bridges; and in the late 1870s Jenkin, with Ewing, investigated the new phonograph.
101. Uncatalogued manuscript in Lord Jenkin of Roding's collection. I have retained the original spelling and syntax, except where indicated.
102. His classmates included Tait, as Jenkin's father noted, and Clerk Maxwell, as Stevenson, op. cit. (1), informs us.
103. He was fluent in Italian and French, and competent in German.
104. Some authorities say the degree was a Master of Arts—this discrepancy remains to be sorted out.
105. By now having worked with Liddell and Gordon of London he entered into a partnership with Forde. See entry in *DNB*, 733.
106. W. Thomson, 'Note on the contributions of Fleeming Jenkin to electrical and engineering science', in Colvin and Ewing, op. cit. (1), clv.
107. Jenkin was a junior partner, taking a quarter of the profits. Letter Jenkin to Thomson, 1865. UGL, Kelvin Collection, J52.
108. For example, in computing. The early days (1940s) of modern computing saw the involvement of some remarkable people. For a study of one of these, and for an appreciation of others, see A. Hodges, *The Enigma of Intelligence*, (London, 1983).
109. C.A. Hampstead, 'Kelvin, instrumentation and the first Atlantic telegraph', Paper 7 in *Papers Presented to the 14th IEE Weekend Meeting on the History of Electrical Engineering, Edinburgh, 4–6 July, 1986*, (London 1987).
110. In the mid-1950s I was employed in the Research and Development

Laboratories of De Havilland Propellors Ltd, of Hatfield, Herts., England. We were designing an infra-red guided and fused air-to-air missile. At that time there was very little experience of missiles, at least in the UK—there was little knowledge and there were no standard designs. When I left the firm in 1960 there were no 'rule books'. In a recent conversation with Mr Peter Steer, late of British Aerospace, now retired, he informed me that there are now the missile equivalents of design handbooks.

It is also worth while to note that in aircraft design it is possible to make use of the NASA handbook of aerofoils, which contains large numbers of feasible, and proved, forms of wings, etc.

111. After about 1870, while Jenkin and Thomson maintained a consulting interest in telegraphy they made increasingly fewer patent applications, and published less and less on the subject. As we shall see below, Jenkin turned to other things.

112. This was true of the submarine telegraphs, but was also true of other innovations. In the early 1950s air-to-air missiles used valves (vacuum tubes) in their electronics. The decision was taken to use transistors, the 'new' technology; however, the germanium transistors then easily available could not operate at the high temperatures involved, so silicon devices had to be used. Even these could not meet the harsh conditions of missile operations, high temperatures, very large *g* forces and so on. Special testing procedures had to be adopted, resulting in large reject rates and high costs. One silicon transistor effectively cost an average week's wage.

113. The Red Sea cable was a vital part of the link between London and India; with the Indian Mutiny so recently in memory, an effective, reliable, controlled and safe telegraph line was considered to be essential. It was a great disappointment when the existing technology could not meet the desire.

114. *The Report of the Joint Committee appointed by the Lords of the Committee of Privy Council for Trade and the Atlantic Telegraph Company to inquire into the Construction of Submarine Telegraph Cables*, (London, 1861).

115. C.A. Hempstead, 'The early years of oceanic telegraphy: technology, science and politics', *IEE Proceedings*, 1989, 136: 297–305.

116. The Appendices contain valuable and detailed accounts of the many researches that were carried out, and illustrate the interplay between these three factors.

117. Included in this group were Charles Bright, Latimer Clark, Henry Forde, Fleeming Jenkin, Robert Newall, Willoughby Smith, William Thomson, and Samuel and Cromwell Varley.

118. There was much contemporary interest in the Atlantic cables. For the best account, although of the 1865 attempt rather than the completely successful 1866 expedition, see W. Russell, *The Atlantic Telegraph*, (London, 1865).

119. That is, until the introduction of repeater amplifiers.

120. See, for example, the discussions in J.L. Kieve, *The Electric Telegraph: a Social and Economic History*, (Newton Abbot, 1973).

121. R.J. Cain, *Cables in the British Empire*, (Unpublished PhD Thesis, Duke University, 1971).

122. See B. Hunt, ' "Practice vs. Theory". The British electrical debate, 1888–1891', *Isis*, 1983, 74: 341–55.

123. Thomson, op. cit. (106), clv.

124. Ibid., clvi.

125. It is not possible to be categoric in this respect, for there is no complete bibliography of Jenkin's work. Even the Collected Papers, Colvin and Ewing, (1), above, is incomplete.

126. Categories are difficult to define, and thus some of the electrical engineering might fall into mechanical engineering, and vice versa: I have selected by 'end use'!

127. By this term I mean something like computers twenty years ago, when the technology was in use, but had not yet reached the 'rule book' stage. Or missile technology forty years ago, or aeroplanes seventy years ago, and so on.

128. H.C.F. Jenkin, 'Experimental researches on the transmission of electric signals through submarine cables—Part I. Laws of transmission through various lengths of one cable', *Phil. Trans.*, 1863, 152: 987-1017.

129. See, for example, the discussion in B. Dibner, *The Atlantic Cable*, (New York, 1973); and D. De Cogan, 'Dr. Whitehouse and the 1858 trans-Atlantic Cable', *History of Technology*, 1985, 10: 1-15.

130. Thomson, op. cit. (106), clvi.

131. His published papers, with J.A. Ewing, demonstrate the use of the phonograph in studying speech: his awareness of commercial value is indicated in an 1878 handbill advertising experiences with the phonograph to be had at the University Cricket Bazaar, in the Music Hall on 22 and 23 March 1878. Personal explanations and recordings were overseen by Mrs Jenkin, and could be experienced for a half-crown ($12\frac{1}{2}$p), lectures could be attended for one shilling (5p), and 'Phonograms or specimens of the writing which speaks will be sold for sixpence ($2\frac{1}{2}$p). A copy of the handbill is in Lord Patrick Jenkin's collection, as is a sample of the 'writing'.

132. That is, how to avoid the necessity of running lamps in series—solved, of course, by Edison.

133. See, for example, B. Bowers, *A History of Electric Light and Power*, (London, 1982), Chapter 16, 247-65.

134. For his comments see H.C.F. Jenkin, *Reports on Electrical Instruments*, (London, 1862).

135. See the comments in *Minutes of the Proceedings of the Institution of Civil Engineers*, op. cit. (9).

136. That is, on the 'quality' of the engineers.

137. H.C.F. Jenkin, 'On the practical application of reciprocal figures to the calculation of strains on framework', *Transactions of the Royal Society of Edinburgh*, 1869, 25: 441.

138. Ibid., 441-7.

139. Kelvin Collection, University of Cambridge Library, J25.

140. Ibid.

141. H.C.F. Jenkin, 'On braced arches and suspension bridges', *Transactions of the Royal Scottish Society of Arts*, 1870, 8: 138.

142. *Minutes of the Proceedings of the Institution of Civil Engineers*, op. cit. (9), 370.

143. Colvin and Ewing, op. cit. (1), 2: 271-339. First published in the *Transactions of the Royal Society of Edinburgh*, 1876-8, 28: 1 and 703.

144. Ibid., 271.

145. Ibid.

146. Ibid.

147. Ibid., 279 *et seq.*

148. *Proceedings of the Royal Society of Edinburgh*, 1886, 14: 119.

149. *Proceedings of the Institution of Electrical Engineers*, 1885, 14: 349.

150. Thomson, op. cit. (106), clix.

151. *Minutes of the Proceedings of the Institution of Civil Engineers*, op. cit. (9), 375.

152. That is, University College Library has nothing of Jenkin's listed other than his published work. Departmental archives have not yet been consulted.

153. According to R.M. Birse, *Engineering at Edinburgh University*, (Edinburgh, 1983), 94, there was a Chair of Technology from 1855 to 1859; Jenkin was the first Professor of Engineering. Lord Jenkin of Roding has expressed the opinion that Fleeming Jenkin was originally appointed to the Chair of Technology, but that this title was later changed. I have not yet been able to confirm or refute this claim.

154. Buchanan, op. cit. (3), 171.

155. Ibid.

156. H.C.F. Jenkin, 'Technical education', in Colvin and Ewing, op. cit. (1), 158. First read to the Royal Scottish Society of Arts in Edinburgh, 11 January, 1869.

157. See ibid., 174. Jenkin, like those who tried a similar thing in the late 1960s, found it difficult to define a 'liberal education'.

158. See his criticism of the British system in ibid., 158–65. He was particularly scathing about what he saw as an invidious comparison between the quantity of instruction devoted to the fine arts, rather than to the technical arts.

159. H.C.F. Jenkin, 'On science teaching in laboratories', Colvin and Ewing, op. cit. (1), 183–90. Paper first read at the International Health Exhibition, (London, 1884).

160. Perhaps following Maxwell, but he did not acknowledge any indebtedness. See, for example, the brief discussion of the introduction of laboratories into science teaching in D.S.L. Cardwell, *The Organisation of Science in England*, (London, 1972), 138–40.

161. Jenkin, op. cit. (159), 184.

162. Ibid.

163. Ibid., 185.

164. Ibid.

165. Buchanan, op. cit. (3), 171.

166. Birse, op. cit. (153), 107.

167. See, for example, the discussion of such matters in J.G. Crowther, *The Cavendish Laboratory 1874–1974*, (London, 1974).

168. Buchanan, op. cit. (3), 173.

169. H.C.F. Jenkin, 'On telpherage', *The Electrician*, 1883, 11: 591. This paper was first presented as an introductory address delivered to the Class of Engineering, University of Edinburgh, 30 October, 1883.

170. Ibid.

171. See archives of the Physics Faculty at Cornell. The required texts were European in origin, including French, German and English texts. In the early years the Cornell Faculty looked to England, but later towards the German universities for their inspiration.

172. Perhaps, indeed, as far back as his work concerned with the determination of electrical standards.

173. See, for example, the discussion of his 'Fragment on truth', above.

174. See *Minutes of the Proceedings of the Institution of Civil Engineers*, op. cit. (9), 367; also the numerous letters in the Kelvin collection at the University of Glasgow library, many of which are concerned with business matters.

175. It could, of course, have been the other way round.

176. See letter to William Thomson in which Jenkin asks about Boscovich. University Library, Glasgow, Kelvin Collection, J68. See also discussions in various parts of Smith and Wise, op. cit. (7).

The Sources for a Biography of Oliver Heaviside

A.C. LYNCH

INTRODUCTION

Although Oliver Heaviside died in 1925, no full-length biography of him appeared until 1988. There is still some difficulty in knowing what to believe about him: some of the available information is misleading even when it is true or nearly true. Yet it is important to know about his way of life, because there were features of it that delayed the acceptance of his work for many years and may even have caused some of it to be lost for ever. If a new Heaviside appeared, how should we avoid the same thing happening again?

HIS LIFE

The agreed facts about Heaviside's life are that he was born in Camden Town, then on the outskirts of north London, in 1850, the youngest of four brothers. His father was a wood-engraver. His mother was sister-in-law to Sir Charles Wheatstone, but the connection is less significant than it might appear, for Wheatstone had, in defiance of convention, quietly married his cook. There is no record that Wheatstone helped the Heaviside family, but he probably used his influence to find jobs for three of the boys. Two of them, Arthur and Oliver, became telegraphists with the Great Northern Telegraph Co., which operated a cable between England and Denmark, and both of them did well there; they became interested not only in the day-to-day routine of telegraphy but also in the mathematical theory of it and in experimental improvements in faster transmission and in fault location.

When the Post Office took over the cable in 1870, Arthur went with it and rose eventually to a senior position in the Post Office. But Oliver was ill at the time, and although he recovered in most respects he had become partially deaf and remained so for the rest of his life. He never worked for an employer again. He earned a little by writing technical articles for *The Electrician*, newly started. These articles were abstruse, and after a few years a new editor stopped them for a time. Heaviside's income after 1896 was a Civil List pension of £120 a year (afterwards raised to £220) plus whatever little he could get by writing.

In 1889 Oliver and his parents moved to Devon, and Oliver never crossed the county boundary again. After the deaths of his parents he lived alone except for a housekeeper, and finally he lived completely alone from 1916 to 1925. He tried to keep himself warm by the lavish use of gas fires, and had difficulty in paying his gas bills. He survived one winter with no heating because the gas had been cut off. He died in 1925 at the age of 74.

HEAVISIDE'S WORK

Heaviside is famous chiefly for a conjecture that there was a conducting layer in the upper atmosphere that could reflect radio signals round the curvature of the earth. He ought to be remembered not so much for that as for developing the theory of telegraph and telephone line transmission and for his method of solving differential equations by taking liberties with the notation (this method is still in use). He was influential in the adoption of vectors rather than quaternions, and we still use his notation for writing vectors; he invented a number of useful terms, such as permittivity and inductance; and he was one of the first to advocate rationalized units, though not in the form in which we use them now.

His work on telephone transmission was practically complete by 1890. It was controversial in that his mathematical techniques were unorthodox; it was justifiable only because it led to results in accordance with observation. His later work was on electromagnetic fields rather than lines, especially on the effects of moving charges, and much of it remains unpublished. It is possible that it includes anticipations of important work rediscovered by other mathematicians.

Many of his contemporaries either ignored or denounced his work, probably for three reasons: first, he published chiefly in *The Electrician*, where mathematicians and physicists did not see the work; second, those few who saw it were put off by the formidable and peculiar mathematics; and third, Heaviside himself was inaccessible, never attending scientific meetings. Few people visited him: Oliver Lodge once, Silvanus Thompson once, Fitzgerald twice, Bromwich the mathematician once. G.F.C. Searle's visits were more numerous: in most of the summers from 1892 to 1913, and at Christmas in 1920 and 1922. Searle was sent for in late 1924, found Heaviside gravely ill, and had him taken to a nursing home, where he died a few weeks later.

BIOGRAPHIES

The earliest attempts at biography were of course the obituaries. They described his work and influence but not his early life. Next came a paper by the mathematical historian E.T. Whittaker.[1] This is mainly about Heaviside's work, but there is some account of his early life, based on hints in Heaviside's own statements, which give a false impression of a minimal and miserable education. Heaviside had indeed complained that mathematics was taught badly; he thought that geometry in particular (and probably other branches too) should be taught as a series of experimental discoveries rather than as exercises in logic. This was intended as a criticism of mathematical teaching in general, not—as Whittaker seems to assume—as a complaint about his own teacher. Heaviside's mastery of English and his ability to coin words when necessary are sufficient proof that in some respects at least his education was good.

In 1950 the Institution of Electrical Engineers (IEE) celebrated the centenary of Heaviside's birth.[2] G.F.C. Searle, then aged 86, contributed an account of Heaviside's home life. Another paper about Heaviside's life, by Sir George Lee, is based on information collected, mainly from the Heaviside family, by H.J. Josephs and his assistant Mrs Beryl Turner, who began their enquiries with no knowledge of Searle's existence. They found out a great deal that Searle either did not know or did not think important, especially about Heaviside's education. They were, of course, working twenty-five years after Heaviside's death and they were fortunate to find out as much as they did. (For information about Searle and Josephs, see the Appendix to this paper).

By 1963 Josephs had written a long account of Heaviside's life and work. He says that it is not a biography, merely background information intended to make it unnecessary to refer to the Heaviside manuscripts held by the IEE. Nevertheless, it would have served as a biography if it had ever been published; but it exists only as a typescript in the IEE Archives.

Searle had intended to write a longer account, but failed to find a publisher for it. In 1987 Ivor Catt, who had acquired a 50,000-word draft, published it.[3] It is the whole of what Catt had obtained,[4] but it looks incomplete. In this book, Searle is scornful of Heaviside's ignorance of Greek and misuse of Latin. The bad Latin was probably intentional, but Searle either did not see Heaviside's joke or thought it in bad taste to mishandle a noble language.

The biography published in 1988 is by Paul Nahin,[5] an American who had not seen the Searle book and may not have seen the Josephs draft. It is open to some criticism. Its chapter on the horrors of nineteenth-century London, although defensible, may be misleading: it supports, perhaps unintentionally, Whittaker's picture of Heaviside's miserable upbringing. It plays down the importance of Heaviside's mathematical methods; they may be obsolete now, but they were used to great effect for many years. But this is the best of the biographical work we have, and unlikely to be superseded.

THE SOURCES

The biographies do not explain fully what had made Heaviside's position so isolated. They draw on four sources: (1) Heaviside's own statements in his published papers (but these are very few); (2) his letters, of which upwards of 1,000 are thought to survive; (3) Searle's recollections; and (4) recollections by non-scientific people, including his relatives. These sources conflict to some extent, especially about his education. They also differ about the extent to which he was isolated; he may have received more visitors than are recorded.

The letters, and no other source, reveal the origin of Heaviside's animus against the 'Establishment' of his day. Oliver and Arthur had collaborated in a paper about transmission; since Arthur was a Post Office employee, he needed permission to publish it. The draft was submitted to W.H. Preece (afterwards Sir William), who did not accept (and probably did not understand) a mathematical appendix by Oliver. He wanted it changed but Oliver stood his ground, and permission was refused. From then on, Oliver blamed Preece, rightly or wrongly, for every difficulty he had in getting his work into print. His letters to Lodge are peppered with venomous references to 'the eminent scienticulist' or some other contemptuous description. At the IEE's Heaviside Centenary celebration, this quarrel was mentioned only in passing, in spite of its importance in Heaviside's life. Josephs told me in 1988 that this was an agreed policy of the organizers. It probably accounts for Searle's remark to me after his lecture at the Centenary meeting—'I wasn't allowed to say all I wanted to say: wheels within wheels'—which I did not understand until, in 1987, I read Heaviside's letters to Lodge.

THE NEED FOR MORE STUDY

Besides the problem of Heaviside's hatred of the Establishment, there is his refusal to attend scientific meetings. Josephs thinks he was too poor to buy suitable clothes. Another plausible reason is his deafness, the extent and persistence of which are difficult to assess. He could hold conversations, but of course this would not guarantee that he could hear a speaker in a large hall in the days before sound amplification. Certainly, he chose to cut himself off from his contemporaries.

Although his isolation may have been imposed partly by his poverty and his deafness, it was also caused by his intolerance towards those who failed to understand what he was trying to do. There were other factors too: he thought Preece was his enemy, and because he was self-taught as a mathematician he expressed himself in ways that others could not follow.

His letters are important and revealing, and as a biographical source they have not been fully used. Many of those he received are in the IEE Archives, but those he wrote are dispersed, and it is to be hoped that they can in some way be made more accessible. The IEE is now (late 1990) sponsoring a search for them.

APPENDIX: TWO OF THE PEOPLE IN THE STORY

G.F.C. Searle (1864–1954) was a mathematical and experimental physicist who worked in the Cavendish Laboratory, Cambridge, for sixty years.[6] As a young man he worked with J.J. Thomson in determining the 'ratio of the units'—equivalent in modern terms to measuring the velocity of electromagnetic waves. For over forty years he organized the teaching laboratories for the first two years of the physics course, devising ingenious experiments which introduced a wide range of mechanical, optical and electrical techniques. Having grown up in the traditions of Kelvin and Stokes, he was well able to handle the mathematics of moving charges and currents (the subject that brought him into contact with Heaviside), but he was never interested in Einstein's relativity and always retained his belief in a Kelvin-type mechanical ether.

Although at heart a kindly man with strong religious beliefs, Searle was alarming in manner and sometimes eccentric in conduct. His sense of humour was strong but sometimes savage; he has even been described as a bully.[7] The character of M.H.L. Gay in C.P. Snow's novels seems to borrow a lot from Searle. Searle claimed that he and Heaviside were much alike in character.[8] Both Dr Vigoureux (of the NPL) and I have found ourselves believing him;[9] but it cannot be quite true, for their backgrounds were widely different. Searle was religious, an academic mathematician and a member of the Establishment of his day; Heaviside was agnostic, he used unorthodox mathematical methods and notation, and he withdrew himself from society.

H.J. Josephs was originally trained as a draughtsman, but became interested in telephony through his wartime experiences in 1918. He worked for the Post Office, and in the 1920s he was allocated to the Research Station at Dollis Hill to help W.G. Radley (later Sir Gordon) in his studies of currents induced from overhead power lines into telephone lines, for which they made much use of Heaviside's theories and methods. Josephs retired in 1960 to devote himself to studying Heaviside's unpublished papers. He is still doing so (1990) and writing about Heaviside's discoveries, but he makes no attempt to publish what he writes.

Acknowledgement
Most of this paper appeared in the Proceedings of the IEE Weekend Meeting on the History of Electrical Engineering, Twickenham, 1988.

Notes and References

1. E.T. Whittaker, 'Oliver Heaviside', *Bulletin of Calcutta Mathematical Society*, 1928, 20: 199–200; reprinted in D.H. Moore, *Heaviside Operational Calculus*, (New York, 1971).
2. Various authors, *The Heaviside Centenary Volume*, (London, 1950).
3. G.F.C. Searle, *Oliver Heaviside, the Man*, (St Albans, 1987).
4. I. Catt, personal communication.
5. P.J. Nahin, *Oliver Heaviside, Sage in Solitude*, (New York, 1988).
6. Obituary notices in *Physical Society Year Book*, 1955.
7. Sir Mark Oliphant in conversation with the present author, 1990; tape recording in the IEE Archives.
8. Searle, op. cit. (3).
9. P. Vigoureux in conversation with the present author, 1988; tape recording in the IEE Archives.

Building Thomas Edison's Laboratory at West Orange, New Jersey

A Case Study in Using Craft Knowledge for Technological Invention 1886–1888

W. BERNARD CARLSON

Every year thousands of people visit the Edison National Historic Site in West Orange, New Jersey. There, in the laboratory that Edison occupied for over forty years, they marvel at the large machine shop, the first motion picture studio and the exquisite wood-panelled library. Others come to see where the great man invented, spat on the floor and took his famous catnaps. However, for the more discerning, the West Orange laboratory is more than just a collection of buildings: viewed as a whole, it is an artefact of the invention process. Edison carefully planned his West Orange laboratory, bringing together the knowledge, people and resources that allowed him to invent rapidly and effectively. Thus, if one were to look closely at the layout of the buildings and how they were equipped, one could discover not only what Edison needed to invent, but more importantly how he worked.[1]

This essay examines Edison's resource base by narrating the planning and building of the West Orange laboratory from 1886 to 1888. It begins by reviewing Edison's activities in the mid-1880s and why he chose to build a new laboratory. The next two sections narrate the construction of the West Orange laboratory and provide a detailed description of the buildings and equipment. From this description, I suggest that the design of the laboratory reveals how Edison depended primarily on craft knowledge, as opposed to scientific theory, in order to invent. This analysis of the West Orange laboratory raises several questions about the history of industrial research as well as contemporary policy towards research and development, and these issues are discussed briefly in the conclusion.

EDISON IN THE 1880s AND THE DECISION TO BUILD WEST ORANGE

By the mid-1880s, Edison was America's leading inventor. Over the previous fifteen years, he had developed major improvements in telegraphy, the telephone and the phonograph. Yet Edison's most resounding success, by far, had been his incandescent lighting system. From 1878 to 1882, he had perfected a high-resistance lamp and then followed with a complete system of dynamos, conductors, meters and accessories. Not satisfied with selling isolated systems to individuals and firms, Edison insisted on developing central stations which sold the service of lighting to a broad base of customers and thus helped create the modern utility industry. With these accomplishments, Edison's fame was secure, yet both Edison and his adoring public assumed that still more wondrous inventions would be forthcoming.[2]

Because of a recession in the early 1880s, Edison and his men built few capital-intensive central stations but, as business improved after 1885, Edison began to receive a substantial income from the promotion of these stations. In addition, he and his close associates—Samuel Insull, Charles Batchelor, Francis Upton, Sigmund Bergmann and Edward H. Johnson—had organized several companies to manufacture electrical machinery; as the central station business expanded, Edison came to enjoy a steady income from these companies as well. By October 1886, the Edison companies had combined assets of ten million dollars and had installed fifty-eight central stations worldwide. With this business well-established, Edison was able to delegate the day-to-day responsibility to his capable lieutenants and to contemplate striking out in new directions.[3]

Because his electric lighting affairs had required him to spend much time in New York City, Edison had abandoned his famous laboratory at Menlo Park in 1881 and moved his research and family to Manhattan. For the next several years, Edison maintained his office at 65 Fifth Avenue (the headquarters of the Edison Electric Light Company) and conducted his experiments on the top floor of Bergmann's factory at Seventeenth Street and Avenue B.

Although Edison may have planned to leave New York at some time and return to Menlo Park, two events prevented him from going back. First, Edison's wife, Mary Stilwell, died in the summer of 1884. Upset by her death, Edison may not have been able to face returning to the home they had shared at Menlo Park. Second, shortly after this tragedy, Edison's house and property at Menlo Park were put up for sale in a sheriff's auction. The auction occurred because Edison had failed to pay off two promissory notes dating back to 1874 and a New Jersey court had ordered that the money be raised by the sale of his New Jersey assets. Although Batchelor purchased Edison's property at the auction, this incident was enough to discourage Edison from ever returning to work at Menlo Park.[4]

With Menlo Park no longer available, Edison began thinking about building a new laboratory. As early as January 1886, Edison sketched a new laboratory in one of his notebooks. In this scheme, he portrayed a substantial laboratory in the French *mansard* style then popular for monumental public buildings. To emphasize the importance of his laboratory, Edison envisioned a three-storey building with a grand arched entrance and a tower which ingeniously disguised the chimney. Built around a central courtyard, this building was a closed design, giving it a certain air of privacy and secrecy. In the floor plan Edison included rooms for a bookkeeper, a library, chemicals, a machine shop and, most importantly, a place for private experiments.[5] With their emphasis on how the laboratory would look rather than how it would operate as an 'invention factory', these early sketches suggest that Edison saw his new laboratory as a demonstration of his prestige as an inventor.

A key defect of this *mansard*-style laboratory was that it might not have been sufficiently flexible for Edison's ever-changing projects. Where could he add on new rooms or buildings? Could he rearrange the layout of the rooms to suit new experiments? Perhaps for these reasons, Edison set this design aside and concentrated on other personal and technical matters. Among the other matters to distract Edison from this laboratory scheme was his remarriage. Through 1885, Edison courted Mina Miller, the daughter of Lewis Miller, a successful agricultural implements manufacturer from Akron, Ohio. Edison and Mina were married in February 1886 and spent their honeymoon at Edison's new winter home in Fort Myers, Florida. When they returned from their honeymoon, the newlyweds established their household at Glenmont, a spacious estate located in Llewellyn Park, a planned community in West Orange, New Jersey.

Once settled into Glenmont in late April 1886, Edison threw himself back into his electric lighting work. Because the Edison utility companies were clamouring for a new system which could compete with the Westinghouse alternating current equipment, Edison investigated a number of ways to improve the efficiency and range of his system. While he worked to increase the quality of his incandescent lamp filament, he also experimented extensively with high voltage direct current distribution systems. This work culminated in his municipal system, which operated at a higher voltage (200 instead of 100 V) and thus powered more lights over a larger area.[6]

In June, Edison was forced to turn his attention from this system to the production problems at the Lamp Works in Harrison, New Jersey. For some time, Edison had been dissatisfied with the quality of lamps produced by his associate Upton, and when Upton left for a long rest in Europe, Edison took charge of the factory. For the remainder of 1886, Edison stayed at Harrison, working to get lamp manufacture back on a solid basis. Because he was so busy with the lamp factory, Edison moved his laboratory from New York to Harrison. However, by December Edison had become tired of the day-to-day supervision of the Lamp Works, especially since he was unable to devote much time to inventing. As Batchelor noted in his diary, Edison was 'very anxious about his experiments . . . and wants me to arrange to do them for him'.[7] Consequently, after he caught pneumonia in the last week of December and was forced to stay home to recover, Edison chose not to return to the Lamp Works.

Instead, in the first months of 1887, Edison started building a new laboratory. In all likelihood, he undertook the project because his immediate experience with working in the Lamp Works convinced him that while he could supervise a manufacturing shop, he did not particularly enjoy it. Although Edison could have located his laboratory at one of his factories, he decided that it would be best to pursue his inventions at a new facility. Experience probably told him that while invention and production should be located near each other, they should also be kept separate in order that experimental work did not interrupt long production runs. Supported by funds from the various Edison companies, he intended to use this new facility to concentrate on what he did best, invention and development.

At the new facility, Edison probably expected to devote a significant amount of time to electrical lighting, but he was also ready to pursue new projects. While working at Bergmann's factory and the Lamp Works, Edison had begun experimenting with new inventions, such as the Edison effect bulb, the phonoplex, a pyromagnetic generator, an electric torpedo and a street railway system. Angered by the appearance in 1885 of the graphophone, he was also busy improving the phonograph so that it could be manufactured and sold in quantity. In 1880, Edison had patented a machine for magnetically separating iron ore and he was anxious to perfect this device. With a backlog of ideas, he was undoubtedly confident that he could keep a large laboratory busy and thus generate sufficient income.[8]

BUILDING THE WEST ORANGE LABORATORY

Edison decided to have his new laboratory near his family and new home in Llewellyn Park. With the lab close to his home, Edison could keep his irregular working hours and still occasionally see his new wife and family. In January 1887, Edison had his attorney John Tomlinson purchase for $4,650 a parcel of land amid the open fields at the bottom of the hill on which Glenmont was situated.[9] Just as he had done with Menlo Park, Edison chose a rural site for the laboratory in order to be away from noise, confusion and the distractions of the city. Nevertheless, he wanted to be near enough to New York to be able to draw on the city for materials, skilled workers and the investment capital of Wall Street.

For this new location in West Orange, Edison skipped his 1886 plans for an ornate building and began sketching a series of more practical structures. From the outset, he intended to build a substantial laboratory; on the back of an early sketch he estimated the laboratory would cost over $100,000. Central to his new plans was 'a special or secret part to [the] Machine Shops for special things I want sub rasa [*sic*]'.[10] One of the things that Edison had missed while working in both the Bergmann factory and the Lamp Works was a sense of privacy and in the new laboratory he wanted to be sure to have his own room.

To turn his sketches into reality, Edison called upon his close associate Batchelor to handle the details and hire an architect. Batchelor retained Hudson H. Holly, the New York architect who had designed Glenmont. Working with Edison's and Batchelor's ideas, Holly designed a three-storey red-brick building with an attached powerhouse, together measuring 250 feet × 50 feet. Eventually, the main building became known as Building 5 and the powerhouse was called Building 6.[11]

The foundation for the main building was laid in late May and construction progressed through the summer of 1887. From the outset, there were problems at the construction site. As chief architect, Holly subcontracted the work to masons and carpenters with whom he was friendly. In addition, Edison had his own man on the site watching Holly and the contractors. Before long, accusations were flying, with Edison's inspector claiming that Holly was a 'schemer', conspiring with the masons to defraud Edison. Holly was accused of skimping on the gravel and cement in the foundation and letting the masons do shoddy brickwork. Other workers found they could not get along with the contractor and quit in disgust, pejoratively calling the project a 'slaughter-house job'. Hearing these accusations, Edison visited the site in July. After surveying one of the walls of the foundations and finding it out of plumb by over an inch, Edison became 'mad as hops'. He promptly fired Holly and replaced him with another architect, Joseph Taft. Through the rest of the summer, Edison regularly made inspection trips to the construction site, often accompanied by his wife and friends.[12]

Although the main building was to have nearly 40,000 square feet of workspace, Edison decided in July that it would be insufficient. To supplement it, he commissioned Taft to build four one-storey buildings. Laid out perpendicular to the main building, each was 100 feet long and 25 feet wide. In each, Edison set up a different laboratory or shop: one was for electricity, another for chemistry, a third for woodworking and chemical storage, and the last for mining, metallurgy and blacksmith work. In the addition of these buildings to the complex, it appears that Edison's conception of the new laboratory evolved as the buildings were constructed; as he saw the walls go up, he began to envision the new laboratory based not only on a machine shop but on a host of shops and facilities.[13]

Compared to the main building, these ancillary buildings went up with little difficulty. As all the buildings neared completion in September, Edison again turned to Batchelor to supervise equipping the shops and workrooms. Over the next three months, Batchelor installed the steam engine and boilers, the machine tools and the gas, water and telephone lines. Since this was to be an electrical laboratory, Batchelor made certain that the facility was fully wired; in every room one could tap a 3, 8, 100 or 1200 volt electrical circuit. In addition, Batchelor had cables laid so that the dynamos at the laboratory supplied current to electric lights to Glenmont and other houses in Llewellyn Park. Finally, he had '8 loads of experimental stuff' safely transported from the Harrison Lamp Works to West Orange and distributed through the new lab.[14]

As Batchelor went about installing machinery, Edison busied himself with stocking the laboratory. Believing that 'an experimenter never knows five minutes ahead what he does want', he ordered ample supplies of every conceivable substance. He filled up at least one notebook with a list of tools he wanted. As Edison boasted to newspapermen, he purchased 'everything from an elephant's hide to the eyeballs of a United States Senator', and he had sufficient quantities on hand to cover at least five years of experimentation. Gathering these supplies took time and energy, leading Edison to comment in December, 'I am so busy at present in connection with fitting up my Laboratory that I hardly get time to sleep.'[15]

As the buildings neared completion and the supplies were delivered, it became clear that the cost of the new laboratory would easily exceed Edison's original estimate of $100,000. Account records reveal that Edison spent over $60,000 on construction and $47,000 on plumbing, heating and finishing the interiors, bringing the total for the buildings to $107,000. To stock and equip the laboratory, he invested at least $73,000 in 1887 and 1888. As a result, the total cost for the laboratory by the end of 1888 was $180,000. To cover part of the cost, Edison took out a mortgage for $20,000, but the grand total shocked him, leading him to comment, 'The Lord only knows where I am to get the shekels—Laboratory is going to be an awful pull on me.'[16]

EDISON'S RESOURCE BASE AT WEST ORANGE

Although Edison and his men were already experimenting in December 1887, they spent much of 1888 organizing the laboratory. New tools and materials kept arriving and new rooms were equipped for new projects. In fact, throughout the history of the laboratory, its arrangement was in constant flux, with Edison skilfully adapting it to the inventions under way. However, one can reconstruct the general arrangement of the laboratory when it opened in late 1887 and from this layout one can identify the resource base which Edison assembled in order to invent.

By far the most impressive room in the laboratory was the library, which served as Edison's public office and meeting room. Located at the west end of the main building, the library was often the first—and sometimes the only—room a visitor saw in the laboratory. The room was panelled in dark-stained yellow pine and its north wall was graced by a large clock, a gift to Edison from his employees in 1889. Around the sides of the room ran two balconies, divided, as was the main floor, into alcoves. On the main floor, these alcoves were filled with scientific and engineering journals as well as volumes of patents. On the first and second tiers, there were popular and trade magazines, a wide variety of books and an extensive collection of ores and minerals. For a time, this mineral collection required a curator and Edison had one of his chemists, Dr George Kunz, sort and label specimens.[17]

In the centre of the library Edison maintained a rolltop desk for meeting with visitors and looking over his correspondence. Next to Edison's desk was a large conference table. Around this table were held numerous meetings, including those of the directors of many of the Edison companies. Initially, there was only a large potted palm in the centre of this room, leaving most of the space open for the display of new inventions and products. Several of the other alcoves on the main floor contained desks used by Edison's secretaries but the most famous alcove contained a cot. Tradition has it that Mrs Edison insisted that this cot be installed because she was tired of hearing that her famous husband was taking his catnaps on the floor.[18]

Across the hall from the library was the main stockroom. It was here that Edison stored the vast range of materials and tools that he felt was essential for invention and development. Included were objects from animals, such as skins, hides, hair, fur, feathers, wool, bristles, teeth, bones, hoofs, horns and shells; natural products, such as woods, barks, roots, leaves, nuts, seeds, herbs, gums and grains; all types of oils, textiles, clays and glass; and all known metals in the form of rods, sheets and tubes. With such a range of materials at hand, Edison eliminated delays arising from having to wait for deliveries and was thus able to speed up invention and development. One experimenter, Reginald Fessenden, recalled that he could always find what he wanted in the stockroom, including an emergency snack; when trapped at the laboratory during a blizzard in 1888, he and Jonas W. Aylsworth, the laboratory's chief chemist, raided the stockroom and dined on buckwheat cakes, maple syrup, dried beef, macaroni with olive oil, dried fruits and coffee.[19]

Beyond the stockroom were the heavy machine shop and engine room, which together occupied about half of the ground floor of the main building. Initially presided over by Batchelor, this shop contained lathes, drill presses and milling machines. With this equipment, Edison and his men could produce dynamos, ore separators, street railway motors and the production machinery necessary for the phonograph and storage battery plants. The machine tools in the shop were connected by overhead shafting to a forty-horsepower Brown steam engine in the one-storey wing on the east end of the building. Also located in this engine room was an Armington & Sims high-speed engine, which drove four electrical dynamos.[20]

Immediately above the heavy machine shop on the first floor was another machine shop known as the precision room. Filled with smaller machine tools, this shop was staffed by highly skilled mechanics who fashioned experimental models and instruments. Supervising the room was John Ott, who served as Edison's personal machinist. The room was particularly important because many inventions began with a small experimental model fashioned in this shop. The significance of the room is underlined by the fact that the main record of experiments was kept here; when an experiment or project was started, it was assigned a number in the 'Precision Room Book', which was subsequently used to track the expenses related to it.[21]

The remainder of the first floor was divided into several rooms used for experiments. One room was fitted up with the glass-blowing equipment and vacuum pumps necessary for making incandescent lamps and Edison regularly drew upon this room in his continuing work to improve his lamp. Another was used by Fred Ott (John Ott's brother) to perform delicate experiments for Edison. Of the remaining rooms, William K.L. Dickson briefly occupied one for his photographic work and another was initially set aside for the assembly of Edison's talking doll.

Of the first-floor experiment rooms, the most significant was Number 12, located just to the right of the stairs running up from the library hallway and on the left side of the first-floor corridor. Initially, this room was intended to be used for drafting but since Edison did not generally use detailed drawings in the course of inventing, the room was soon put to another use: it served as the private workroom Edison had dreamed of having at West Orange. Here Edison conducted his own experiments and tests, working occasionally with an assistant but at other times in solitude. By being just down the hall from the Precision Room, he had immediate access to the tools and craftsmen he needed. Room 12 was plain, equipped with only a few chairs, a large worktable and shelves filled with chemicals and apparatus related to the project at hand. There were probably always a few laboratory notebooks on the table as well as some ordinary yellow tablets, both of which Edison used to sketch and describe his ideas.[22]

An early notebook sketch suggests that Edison intended to have his chemical laboratory on the second floor. To support this laboratory, Edison planned to include a glassblower, lapidary, vacuum pumps, jeweller and translator on this floor. However, once he decided to erect a separate building for the chemistry laboratory (see below), Edison was free to utilize the second floor for other purposes. On the west end of the building over the library, Edison set aside a large room for lectures and presentations. Equipped with a magic lantern projector, the room may have been intended to be suitable for introducing his inventions to investors and the general public. Soon, however, this room became the phonograph recording studio or Music Room. In the early 1890s, Edison temporarily closed the Music Room and instead used this space as a drafting room for planning his large ore milling plant at Ogdensburg, New Jersey. Elsewhere on the second floor, Edison installed a lamp test room where his assistants tested the quality and longevity of new incandescent bulbs before they went into production.[23]

Beyond these few rooms, most of the second floor appears to have been undeveloped space during the 1880s and 1890s. Through the middle of this floor, a hallway was created by lining up cabinets filled with old experimental models and scientific apparatus and the rooms on either side were left vacant, to be utilized as need arose. Edison took advantage of this situation and created offices and workrooms as he undertook new projects. For instance, when he began working on the phonograph in 1888, rooms were set aside for Aylsworth to work on a new cylinder wax and for W.E. Gladstone to develop a better primary battery to power the phonograph's electric motor.[24] Often, new rooms were created on this floor by the erection of thin wooden walls. Thus, while this floor began by being empty, it came to be a vital part of Edison's research strategy by providing space which could be applied to new inventions as they were being developed.

Turning to the smaller laboratory buildings, the one nearest the main road was Building 1 or the Galvanometer Room. Intended to be used for delicate electrical and physics experiments, this building was constructed without any iron or steel hardware which could magnetically interfere with sensitive electrical instruments. Furthermore, it included a series of solid brick piers or tables on which experiments could be conducted without fear of disruptive vibrations. Extensively equipped with galvanometers, electrometers, photometers and Wheatstone bridges, the Galvanometer Room was used by Arthur E. Kennelly and others to perform experiments until 1892. In February of that year, the building was rendered useless since the electrical trolley operated by Newark Passenger Railway Company set up an electric field which threw off the precision instruments.[25]

Building 2, located on the other side of the laboratory yard, was the Chemistry Laboratory. Edison planned this building to have separate rooms for precision balances, the spectrometer and analysis, but instead it was laid out with one main experimental room in the front and a small office in the back.[26] In the main room Edison often kept several chemists at work at different benches, while he preferred to work in the back room on his own.

Perhaps because of the number of chemical experiments Edison had going forward at any given time, Building 2 was reserved for active experiments and the front half of Building 3 (located next door) was used for storing chemicals and apparatus. In the back half of this building, a woodworking shop produced cabinets and the wood patterns necessary for manufacturing different inventions.[27]

While Building 3 was a support facility, Building 4 was used for active research in metallurgy and mining. Here Edison took up his ore milling inventions and developed them from the test model to the pilot plant stage. For this purpose, Edison had stamp mills, rock crushers and several assay furnaces installed in this building. For a time, this building also had the equipment necessary for producing castings and doing other foundry work. The building was generally crowded, filled with kegs and boxes of ore samples which Edison collected in the course of his mining scheme; while most of the samples were iron, he also had numerous specimens of gold and nickel ores. Notably, once Edison moved beyond his ore milling venture of the 1890s, this building became open space which he used flexibly, applying it to different projects as the need arose.[28]

In addition to these buildings, Edison's staff periodically erected additional wooden structures within the laboratory complex. These extra buildings were designed for particular inventions; good examples of such structures were the photographic studio in which Dickson experimented with the kinetoscope and the Black Maria, which was designed especially as a motion picture studio.[29] Furthermore, some experimental work was conducted in the Phonograph Works. Until 1907, these works consisted of two one-storey buildings parallel to Buildings 1 to 4, just beyond the east end of the main building. Since they were equipped with an extensive machine shop, Edison sometimes had parts of his larger inventions (such as his ore separators) made at the Phonograph Works. Because Edison regularly depended on the Phonograph Works to perform certain jobs related to developing his inventions, they too should be considered a part of the West Orange laboratory complex as it existed before 1900.

To utilize the buildings, equipment and resources of the laboratory,

Edison assembled an impressive staff. This staff was highly diverse, and one West Orange researcher recalled that it included 'learned men, cranks, enthusiasts, plain "muckers" and absolutely insane men'.[30] Yet this versatile collection contained all the human elements Edison needed to invent. Most notably, the staff included scientifically trained men such as an electrical physicist (Kennelly), chemists who had earned PhDs in Germany, and several Americans who had studied science in college (e.g. Reginald Fessenden and David Marshall). Supplementing the scientists, Edison employed a number of highly skilled men who worked as experimenters and were dubbed the 'muckers'. Often without formal training, these men possessed extensive hands-on experience with machines and electricity, and Edison could count on them to do much independent creative work. Among the more prominent of the experimenters were Edison's long-time associate Charles Batchelor and William K.L. Dickson, a photographer who helped perfect motion pictures.[31] After the 'muckers' came the skilled craftsmen, whom Edison employed in large numbers: machinists, glassblowers and instrument-makers. Rounding out the lab staff were accountants, clerks and errand boys. Altogether, the staff varied from 50 to 100 men, depending on the number of projects Edison had underway at any given time.

From this description of the West Orange laboratory and its staff, one can see that Edison drew together six key elements to form a resource base for invention and development. First, Edison made sure that he had access to information. In the library, he maintained runs of technical journals and books, trade magazines, patents, and even a mineral collection. To search and utilize these 'databases' effectively, Edison retained technical personnel who could extract the specific information needed for a project. Second, Edison had ample raw materials in the form of natural products, metals and chemicals. Kept in the main stockroom and the chemical storeroom, this stockpile suggests the importance Edison placed on having materials on hand and also his sensitivity to the role materials could play in perfecting an invention. Third, West Orange was well equipped with tools. With two large machine shops, a woodworking shop, a blacksmith shop and a foundry, Edison could develop his inventions rapidly. Fourth, to complement the tools, Edison had in the Galvanometer Room and Chemistry Laboratory a wide range of scientific instruments, permitting him to make precise observations of the phenomena related to his inventions. Fifth, while not always recognized, Edison made certain that he had work space available in the laboratory. As inventions moved from the laboratory bench to the pilot plant stage, Edison had room at West Orange—on the second floor, in the yard, and in Building 4—to scale up and test inventions. Sixth and finally, to use these physical resources Edison brought together the necessary human resources in the form of scientists, experimenters and skilled workmen. In his staff, Edison had a reservoir of talent that could be directed toward a variety of projects and inventions.

ANALYSING THE LABORATORY IN TERMS OF CRAFT KNOWLEDGE

In studying the resources and layout of the West Orange laboratory, one can learn much about how Edison pursued invention. Although one can tease out from the buildings of West Orange a host of clues to his approach to invention, one of the most revealing is Edison's orientation to craft knowledge.

In thinking about craft knowledge at West Orange, I wish to draw attention away from the scientific knowledge that is sometimes emphasized. Although it is easy to focus on the Galvanometer Room and Chemistry Lab and to recall that Edison hired professional scientists, one should be careful about coming to any facile conclusions about the role of science at West Orange. These rooms and these men were only part of the whole laboratory and were not necessarily its centre. Even though it is tempting for twentieth-century visitors to conclude that the Edison lab functioned like a contemporary R&D laboratory—applying basic research to practical problems—such a conclusion does not do justice to the complexity of West Orange. Instead, viewed as a whole, the West Orange laboratory suggests that Edison did not base invention as much on science as he did on craft knowledge.

Since the concept of craft knowledge is not as familiar as scientific knowledge, it is perhaps appropriate to digress and characterize it. Craft knowledge refers to knowledge that is learned by doing, by having immediate hands-on experience. It is based on skill and not on abstract theory or thought. Often craft knowledge may have a strong visual component, with drawings, models and objects playing an important part in the creation and application of this knowledge. Craft knowledge may be verbalized in terms of 'rules of thumb', recipes and descriptions, but it is different from scientific or engineering knowledge in that it is generally neither formalized into theories or laws nor expressed in mathematical terms. Rather than being preserved in books, it is more frequently embodied in the hands and minds of skilled workmen. It is transferred and perpetuated by having workmen move from one shop to another or by training apprentices. One should not assume that craft knowledge is merely the manual techniques used in working materials; instead, it is a complex set of ideas covering problem-solving, aesthetics, the pace of work, the organization of the workplace, customer relations and the relationship between the craftsman, his work and society. Perhaps one of the best pithy descriptions of craft knowledge is provided by A.R. Hall, who has suggested that it is more concerned with knowing how to do something than with knowing why.[32]

Turning back to Edison, I believe that he approached invention not as a scientist but as an inventor whose success was based on craft knowledge. From the beginning of his career as an inventor, Edison had developed new telegraph devices by drawing on his own intimate knowledge of operating telegraph apparatus. First-hand experience with electrical circuits, coupled with new experiments, was the knowledge on which Edison created his new incandescent lighting system. In building West Orange,

Edison brought together a set of resources that would permit him to continue to utilize craft knowledge. In particular, at West Orange he pursued invention by using three aspects of craft knowledge—a variety of specialized techniques, a sensitivity to materials and close observation of phenomena.

At West Orange, Edison drew upon a wide variety of crafts, skills and techniques embodied in both the shops and personnel. By working with machinists, glassblowers and instrument-makers, Edison gleaned new insights into how things were made and how they could be made better. Edison's experimenters or muckers were often machinists or other craftsmen who showed a propensity for using their manual skills in new and creative ways. They had both the ability to see new combinations of objects and the skill needed to implement these combinations. Furthermore, I would suggest that Edison employed scientists not for their theoretical or methodological abilities but because of their ability to work with chemical apparatus and scientific instruments. In conducting research on electricity, Kennelly had clearly acquired a great deal of skill in manipulating electrical apparatus and instruments; for Edison he was valuable because he could provide techniques for handling electricity in much the same way that the machinist provided insight into how metal could be worked and shaped. In the course of invention, Edison drew on the skills and techniques of all these men and combined them with his own manual skills to create new products and processes.[33]

A second aspect to Edison's craft orientation was his sensitivity to materials. Like any highly skilled craftsman, Edison was aware that the right material was often the key to an invention and consequently he studied all types of chemicals, minerals and animal products. Before the building of West Orange, Edison had succeeded with several inventions because he identified and then developed suitable materials. With both the microphone for the telephone and the incandescent lamp filament, Edison found that his knowledge of carbon was the secret of success. At West Orange, Edison emphasized his awareness of the proper material by including the extensive stockroom, the chemistry lab and the metallurgical lab. With these facilities, Edison possessed not only a wide selection of raw materials but also the men and equipment necessary for analysing them. Over the years, the choice of the right material often led to inventions at West Orange. Edison's chemists contributed to the steady improvement of the phonograph by developing better waxes and resins for record cylinders and discs, and the storage battery depended on knowledge of how to use finely powdered nickel as one of the electrodes. Significantly, although Edison turned to his chemists to analyse some materials, he depended on first-hand examination of materials to guide his selection. In developing a tungsten phonograph needle, for instance, Edison and one of his assistants poured a tungsten solution into a series of dishes and then added different substances which would form crystalline structures. Edison studied these crystals under the microscope and selected the one that suited his purposes.[34] Edison's knowledge of materials was extensive and intimate, but it was based on his experience and not on scientific theory.

As the search for a tungsten phonograph needle suggests, a third element in Edison's craft approach to invention was close observation of phenomena. As befits a conscientious workman, Edison carefully studied processes and phenomena, always with an eye for spotting slight variations which he might exploit. To facilitate these observations, he included an array of scientific instruments at West Orange. By using microscopes, galvanometers, precision scales and photographic equipment, Edison investigated the phenomena related to his inventions. With these instruments, he was not a scientist concerned with precise measurements to verify a theory; rather than worrying about why something happened, he wanted to observe how things changed and to use this knowledge to guide the design of new technology. One indication of how Edison employed scientific instruments in the invention process can be seen in the frustration he felt while trying to determine why his storage battery lost its capacity to hold a charge; as he observed, 'In phonographic work we can use our ears and our eyes aided by powerful microscopes; but in the battery our difficulties cannot be seen or heard, but must be observed by our mind's eye.'[35] For Edison the craftsman, invention was a tactile and visual activity, and scientific instruments were extensions of his senses.

In emphasizing the role of craft knowledge at the West Orange lab, I do not wish to suggest that Edison had no appreciation of scientific knowledge. As noted above, he depended on scientists to provide chemical analyses and operate laboratory instruments. Like many other nineteenth-century scientists and engineers, Edison shared an enthusiasm for ascertaining facts through a series of controlled experiments; as a young man, he had absorbed a Baconian empiricism by reading Michael Faraday's *Experimental Researches*. Although he often made fun of the methods used by mathematicians and physicists, Edison appreciated and utilized the results of their calculations. Just as he had depended on Francis Upton to perform crucial calculations on his incandescent lighting system at Menlo Park, so at West Orange Edison looked to Kennelly to provide mathematical analyses related to new electrical generators, meters and distribution networks.[36]

The point I wish to make about Edison and craft knowledge is that scientific knowledge did not predominate at West Orange. Theory and a rigorous methodology formed only a portion of Edison's resource base at West Orange. Rather, the equipment, personnel and layout of the laboratory suggest that craft knowledge was at the centre of the West Orange laboratory. Craft knowledge provided the skills, materials and ideas which Edison manipulated in the invention process. Even more, it supplied him with a broad framework by which he could organize and manage the laboratory. By conceiving and then operating the laboratory as a collection of craft shops—each with its own master craftsmen and special skill—Edison was able to draw on his own experience as a workman and machine shop owner in order to define problems, assign tasks, motivate workers and generally produce inventions. Thus, when Edison said 'I can hire mathematicians, but they can't hire me,' it would be easy to conclude (as many have) that he had no use for mathematics and scientific theory.[37]

However, I would argue that what Edison meant was that mathematics (and implicitly theoretical science) failed to provide him with the insights and cognitive framework he needed to invent. Instead, he found that craft knowledge provided him not only with a host of technical solutions (in manual skills and materials) but also with a way to define problems which hired mathematicians could then solve.

CONCLUSION

Edison's West Orange laboratory is best interpreted as a facility designed to utilize craft knowledge to develop new technology. By seeing invention as a craft-oriented activity, we are able to appreciate why Edison had a mix of artisans and scientists, a wide range of shops and laboratories, large stocks of raw materials, and scientific instruments. To be sure, science played a part in the operation of West Orange, but it provided only a portion of the resource base and it was not the guiding force behind the enterprise.

More broadly, by seeing West Orange from the perspective of craft knowledge, historians may need to rethink the history of industrial research. Previously, scholars have tended to assume that invention was only integrated into the strategy and structure of American business after 1900, when a few large firms such as General Electric, AT&T and DuPont hired professional scientists and established in-house laboratories. The prevalent view has been that the institutionalization of invention is closely linked with using science as a source of new technology. Invention was only rendered reliable and manageable by basing it on science, and so it is assumed that the key change is the creation of in-house labs staffed by scientists.[38]

In contrast, Edison's lab points to a different pattern. The case of West Orange clearly shows that while Edison mastered the process of invention by creating a large laboratory, he did not base his laboratory on science but on craft knowledge. Edison creatively managed craft knowledge so as to render the uncertain process of invention into a dependable element of business strategy and he succeeded in building a profitable industrial empire around his laboratory. West Orange reveals that there was no inherent, deterministic link between the institutionalization of invention and the use of science as a source of new technology; Edison demonstrates that it was possible to institutionalize invention successfully while using another form of knowledge.

This situation raises important questions about why managers and businessmen chose to hire scientists to produce new technology in the early twentieth century; rather than their being a better source of new technology (as is frequently assumed) was it that the values associated with science (predictability and objectivity) appealed to managers and businessmen? Historians might use West Orange as a point of departure for asking new questions about the process by which technology was integrated into modern American business practice.

Along with raising questions about how we interpret the history of research and development, West Orange also provokes questions about our common assumptions concerning the nature of technological change. It is a popular and persistent myth that new technology springs from science. Scientists generate new theories, and engineers simply apply those theories to existing needs. Guided by this model, both American business and the US federal government invest millions of dollars faithfully in pure science. Yet this conception may overlook an important aspect of the creation of new technology, namely the role of hands-on, tactile, craft knowledge. How much new technology today is based on this sort of knowledge? To be sure, science has led technologists to insights and breakthroughs that would have not been possible without theory, but technological creativity still consists of knowing how to transmute theory into a working machine. Again, as one looks at West Orange, one sees that Edison's strength lay in his ability in knowing how things worked. Thus, rather than seeing Edison as an outmoded figure of the nineteenth century, perhaps we should celebrate him as a hero whose approach to invention is just as revealing and insightful today as it was when he first opened the West Orange laboratory one hundred years ago.

Acknowledgements

This paper is based on research conducted under contract for the National Park Service and I am grateful for their support. I conducted this research jointly with A.J. Millard of Bentley College. I wish to thank Edward J. Pershey, Mary B. Bowling and Eric Olsen of the Edison National Historic Site for their assistance in working with the Edison archives. Earlier drafts of this paper were read by Harry M. Collins, Reese V. Jenkins, Bryan Pfaffenberger and Darwin Stapleton and I have benefited from their comments and encouragement. Naturally, this paper reflects my opinions and interpretation and not necessarily those of the National Park Service, Dr Millard or my colleagues. A version of this paper was presented at Industrial Research at Edison's West Orange Laboratory: A Centennial Symposium, Edison National Historic Site, West Orange, New Jersey, 25 April 1987.

Notes and References

The following abbreviations are used in these notes: ENHS, Archives at the Edison National Historical Site, West Orange, New Jersey; PN, with year/month/day, Pocket Notebook; N, with year/month/day, Notebook; TAE, Thomas A. Edison; WO Lab, West Orange Laboratory.

1. For a general discussion of resources or factor endowment as it relates to technological activity, see Nathan Rosenberg, *Technology and American Economic Growth*, (White Plains, NY, 1977). For the concept of method or style of invention—particularly as it relates to Edison—see Thomas P. Hughes, 'Edison's method', in W.B. Pickett (ed.), *Technology at the Turning Point*, (San Francisco, 1977) and Thomas P. Hughes, *Networks of Power: Electrification in Western Society*, (Baltimore, MD, 1983), 18–46.

2. Details of Edison's activities during the 1880s may be found in his several biographies; see Frank L. Dyer and Thomas C. Martin, *Edison: His Life and Inventions*, 2 vols, (New York, 1910); Matthew Josephson, *Edison: A Biography,* (New York, 1959); Thomas P. Hughes, *Thomas Edison: Professional Inventor*, (London, 1976); and Robert Conot, *A Streak of Luck*, (New York, 1979).

3. Josephson, op. cit. (2), 300–1.

4. Conot, op. cit. (2), 218–19.
5. N851001, 62–68, about January 1886, ENHS. I am grateful to Paul Israel for bringing these sketches to my attention.
6. On Edison's research into high-voltage systems, see 'A new Edison system of distribution', *Electrical World*, 1887, 10: 42. On the Edison municipal system, see J.J. Vail to TAE, 22 May 1886, 1886 Electric Light, Edison Co. for Isolated Lighting File, ENHS; TAE to John Ott, 16 May 1887, 1887 Electric Light, General File, ENHS; 'The new Edison municipal lamp', *Electrical World*, 1888, 11: 74; TAE, Municipal lamp cutout caveat, 23 May 1889, N(880601).2; and TAE, Sketches for municipal lamp, N890001.
7. Edison's activities in 1886 are summarized in Charles Batchelor's Journal, 1886–7, Cat. 1336, ENHS (hereafter cited as Batchelor Journal, 1336). Quote is from this journal, entry for 1 December 1886.
8. TAE, [List of Projects], 30 December 1887, 1887 Phonograph General File and TAE notes, N871210.2, ENHS.
9. J. Tomlinson to TAE, 1 February 1887, 1887 WO Lab; Voucher 579, 1 December 1887, 1887 Voucher Series, ENHS.
10. Early cost estimates are from TAE note and sketch, n.d., and quote is from TAE to C. Batchelor, 6 April 1887, both in 1887 WO Lab File, ENHS.
11. Batchelor Journal, 1336, entries for 16 April, 2, 3, 9, and 19 May 1887; the building specifications submitted by Holly are in 1887 WO Lab, Specifications File, ENHS.
12. Jeff Waldron [construction inspector], PN870716; Taft's building specifications are in 1887 WO Lab File, ENHS.
13. Entry for 15 August 1887, PN870716, ENHS.
14. Batchelor Journal, 1336, entry for 5 September 1887; C. Batchelor Journal, Cat. 1337, entries for 19 September, 7 October, 10 and 25 November, 1 and 23 December 1887. For Batchelor's notes on fitting out the lab, see N870600. On installing the gas and water mains, see Vouchers, 339, 347, 648, 649 and 653, 1887 Voucher Series, ENHS.
15. First quote is from TAE note, 24 October 1887, 1887 WO Lab File; second quote is from R.E. Eaton to TAE, December 1887, 1887 Ore Milling File; third quote is from TAE to R. Brigg, 21 December 1887, 1887 WO Lab File, ENHS. For a sample of the exotic materials Edison ordered, see Otto Gerdau to TAE, Bill, 29 December 1887, Voucher 759, 1887 Voucher Series, ENHS.
16. Cost figures are from TAE notes, n.d., 1887 WO Lab File and N.R. Speiden, 'Historic Site Report for Building 1'; quote is from TAE note, 6 July 1887, 1887 Edison Electric Light Company File, ENHS. Edison probably partly offset the heavy expenses of the lab by signing several promissory notes for $23 000 in November 1887 with the Edison Lamp Company and Edison Electric Light Company; see F.X. Hastings to TAE, 3 December 1887, Voucher 571, 1887 Voucher Series, ENHS.
17. W.K.L. and Antonia Dickson, *The Life and Inventions of Thomas Alva Edison*, (New York, 1894), 283; and Reginald Fessenden, 'The inventions of Reginald A. Fessenden', Part VIII, *Radio News* (August 1925) in Fessenden Biographical File, ENHS.
18. Dyer and Martin, op. cit. (2), 642–5.
19. Fessenden, op. cit. (17), Part VIII.
20. 'Edison's new laboratory', *Scientific American*, 1887, 57: 184 (hereafter cited as *Scientific American*, 57; Dickson, op. cit. (17), 293; and Dyer and Martin, op. cit. (2), 647–8.

21. *Scientific American*, 57; and John F. Ott, 'Precision room book', N880127, ENHS.
22. Dyer and Martin, op. cit. (2), 648–50.
23. TAE sketch, n.d., N870000.3; *Scientific American*, 57; TAE to Garrison, 21 Dec. 1887, 1887 WO Lab File, ENHS.
24. *Scientific American*, 57; Fessenden, op. cit. (17), Part VIII; Dickson, op. cit. (17), 295.
25. 'Historic site report for building 1', and David Lawrence Pierson, *History of the Oranges to 1921*, 3 vols, (New York, 1922), 2: 428.
26. *Scientific American*, 57.
27. Dyer and Martin, op. cit. (2), 655.
28. Dickson, op. cit. (17), 323.
29. On the photographic building and the Black Maria, see W.K.L. Dickson, 'A brief history of the kinetograph, the kinetoscope, and the kineto-phonograph', in *A Technological History of Motion Pictures and Television*, Raymond Fielding (ed.), (Berkeley, 1980), 9–16. This article was originally published in *Journal of the Society of Motion Picture Engineers*, 1933, 21. See also Gordon Hendricks, *The Edison Motion Picture Myth*, (Berkeley, 1961).
30. David Trumbull Marshall, *Recollections of Edison*, (Boston, n.d.), 60.
31. These two experimenters have been studied extensively; see Walter L. Welch, *Charles Batchelor: Edison's Chief Partner*, (Syracuse, 1972), and Hendricks, op. cit. (29).
32. To the best of my knowledge, no historian of technology has developed an explicit definition of craft knowledge. Instead, it has been discussed under a number of different terms including technical skill, shop culture and craft tradition. In thinking about Edison and craft knowledge, I have been influenced by the following studies: A. Rupert Hall, 'On knowing, and knowing how to . . .', *History of Technology*, 1978, 3: 91–103; Merritt Roe Smith, *Harpers Ferry Armory and the New Technology*, (Ithaca, 1977), 62–8; Monte A. Calvert, *The Mechanical Engineer in America*, 1830–1910, (Baltimore, 1967), 6–8; John M. Staudenmaier, *Technology's Storytellers*, (Cambridge, MA, 1985), 114–20; Walter G. Vincenti, 'Technological knowledge without science: the innovation of flush riveting in American airplanes, *ca.* 1930–*ca.* 1950', *Technology and Culture*, 1984, 25: 540–76; Brooke Hindle, *Emulation and Invention*, (New York, 1983); and James R. Blackaby, 'How the workbench changed the nature of work', *American Heritage of Invention and Technology*, 1986, 2: 26–30.
33. Edison's reliance on skilled craftsmen is discussed in a sociological report prepared for the National Research Council in 1927. See Maurice Holland, 'Edison—organizer or genius?', Box 12, Fol. 7, Edison Papers, Archives and Library, Edison Institute, Dearborn.
34. On Edison's interest in materials, see Byron M. Vanderbilt, *Thomas Edison, Chemist*, (Washington, DC, 1971). On the role of carbon in the telephone and incandescent light, see Robert Friedel and Paul Israel, *Edison's Electric Light: Biography of an Invention*, (New Brunswick, 1986), 93–114. On the development of the alkaline storage battery, consult Richard H. Schallenberg, *Bottled Energy: Electrical Engineering and the Evolution of Chemical Energy Storage*, (Philadelphia, 1982), 350–73. For the example of the tungsten needles, see Holland, op. cit. (33).
35. 'Edison's alkaline storage battery', Meadowcroft Box 73, ENHS.
36. On the ways in which science influenced technology in the nineteenth century, see Hall, op. cit. (32), 101. On the impact Faraday had on Edison, consult Hughes, op. cit. (2), 5 and Josephson, op. cit. (2), 61. On Upton's role in developing the incandescent lamp, see Hughes, op. cit. (1), 33–7. For Kennelly's calculations at West Orange, see his 'Record books of observations, calculations,

and data compiled in the galvanometer building', vols 1 and 2, 20 October 1888 to 28 July 1889 and 28 July 1889 to 24 February 1890.

37. Josephson, op. cit. (2), 357.

38. For an overview of the historical literature on industrial research, see W. Bernard Carlson, 'Industrial research in America: a select guide to historical studies', *History of Science in America: News and Views*, 1984, 3: 2-3. Among the more important studies are Leonard S. Reich, *The Making of American Industrial Research: Science and Business at GE and Bell, 1876-1926*, (Cambridge, 1985); George Wise, *Willis R. Whitney, General Electric, and the Origins of U.S. Industrial Research*, (New York, 1985); Stuart W. Leslie, *Boss Kettering: Wizard of General Motors*, (New York, 1983); Jeffrey Sturchio, 'Chemists in industry: studies in the historical application of science indicators', PhD Dissertation, History and Sociology of Science, University of Pennsylvania, 1981; Reese V. Jenkins, *Images and Enterprise: Technology and the American Photographic Industry, 1839 to 1925*, (Baltimore, 1975). For a valuable critique of this literature, see Michael Aaron Dennis, 'Accounting for research: new histories of corporate laboratories and the social history of American science', *Social Studies of Science*, forthcoming. For an expanded exploration of the significance of Edison's West Orange laboratory in the history of industrial research, see W. Bernard Carlson, 'Heroic inventors and the evolution of industrial research: Thomas A. Edison and the West Orange Laboratory, 1887-1931', Paper presented at the Organization of American Historians, Philadelphia meeting, April 1987.

Edison and Early Electrical Engineering in Britain

BRIAN BOWERS

INTRODUCTION

The name of Thomas Alva Edison is so well known in the context of electrical engineering history that it may be a surprise to find that his period of active involvement with electrical engineering in Britain was quite short.

Born in 1847, Edison was trained as a telegraph operator. His career as an inventor began with improvements in telegraph instruments and his first involvement with British electrical engineering was a visit in 1873 to persuade the British Post Office Telegraph Department to adopt some of his instruments. In 1877 he invented the phonograph, but his interests soon turned to electric lighting. Having developed a fairly satisfactory incandescent filament lamp, he designed generators and all the other equipment necessary for a public electricity supply system. Edison was eager to exploit this in Britain, even before the whole system was working in the USA. In the event his Holborn Viaduct generating station in London began supply to its first customers in January 1882, eight months before his Pearl Street station opened in New York.

Edison's venture into electric lighting in Britain was dominated by his patents clash with Joseph Swan, which led to the merger of the Edison and Swan interests in Britain. The merged company dominated the lamp market, but Edison soon ceased any active involvement with electricity supply in Britain. Nevertheless, Edison remained a 'newsworthy' figure, and his second visit to Britain, in 1889, was reported at length.

TELEGRAPHS

In 1871 Edison's telegraph work was flourishing. He was manufacturing printing telegraphs in quantity and the assured income from those enabled him to work on automatic telegraphy.[1] The automatic telegraph was vital to the telegraph business because it could send messages which had been prepared on punched paper tape very much faster than an operator could work. As many as ten operators could be preparing paper tapes for messages which were sent over a single telegraph line and printed automatically at the receiving end.

From the 1860s the British Post Office used Wheatstone automatic equipment. In 1871 Daniel Craig wrote to Frank Scudamore seeking to interest him in American equipment.[2] Craig, a pioneer of automatic telegraphy in the USA and one of the founders of the Automatic Telegraph Company, had engaged Edison to improve the existing American instruments. Scudamore was second secretary to the British Post Office. Craig said he had instruments which 'can send 30,000 words per hour with absolute correctness', whereas he understood that the Wheatstone machines used by the British were averaging only 3,600 words per hour. He invited Scudamore to send someone to inspect the system. It is not clear whether or not Scudamore took up that offer, but George Harrington, president of the Automatic Telegraph Company, continued to promote Edison's automatic telegraph in Britain.[3] Early in 1873 George Gourard, agent for the Automatic Telegraph Company, met Scudamore in London. As a result the Post Office were persuaded to have a demonstration of the American system, although they had doubts as to whether it would be worth adopting, even if it were as good as claimed, because the existing traffic did not use the Wheatstone system to its full capacity. It was agreed that if the American system achieved 500 words per minute over a 300-mile line then the Post Office would pay the expenses of the demonstration and consider adopting it.[4]

Edison left the USA for England on 23 April 1873. It was his first overseas visit, and he was accompanied by Jack Wright of the Automatic Telegraph Company. The trials took place between London and Liverpool in late May. They quickly found that the American apparatus would not work at speed with underground lines, but on the overhead line between the edge of London and the edge of Liverpool they succeeded in reaching the required speed of operation. The Post Office, however, did not adopt the system. Edison returned to America in June.

PHONOGRAPHS

The phonograph was a mechanical device, but Edison saw it at first as an adjunct to the telephone. It is relevant to include some account of the phonograph in this paper on Edison and electrical engineering because it greatly enhanced Edison's reputation.

In 1877 Edison was thinking about the problem of training telegraphists to understand Morse code at speed. It occurred to him that if the Morse signal could be recorded and played again at a lower speed it would help the learners. There were already instruments available that recorded Morse messages by embossing a paper tape and, as mentioned above, there were instruments that transmitted Morse messages from a prepared paper tape punched with holes to represent the dots and dashes. Ten years earlier Edison had sketched an arrangement with two typical American telegraph office instruments combined to record Morse and replay it more slowly.[5] Later, in some autobiographical notes, Edison wrote:

> I was very ambitious to be able to take press report . . . but it came faster than I could write it down legibly—at this time I conceived the idea of taking two old Morse Registers which recorded the dots & dashes by indenting a continuous strip of paper, the indenting point

> being worked by a lever and magnet. . . . I arranged the second register so that, the strip passing through it, the indentations were made to actuate a delicate double lever causing the Local circuit of a sounder or receiving instrument to be opened & closed corresponding exactly to the original signals. . . . When press was coming over the wire . . . at . . . 40 words per minute . . . the 2nd Register repeated these signals audibly . . . at . . . 25 or 30 words per minute . . . it was this instrument which gave me the idea of the phonograph while working on the Telephone.[6]

Edison was working on the telephone in 1877. The Scot, Alexander Graham Bell, who had emigrated to Canada, had made his first telephone in 1875, but it had a very limited range because its only power came from the speech of the person talking. Edison had improved it by introducing the carbon microphone. With this the speech signals controlled the current from a battery so that more power was available and the sounds could be transmitted over longer distances.

Edward Johnson, an associate of Edison's, lectured on Edison's inventions and gained the strong impression that a method of recording the human voice would appeal to the public; irrespective of any application to the telephone. He wrote about the idea in a letter to *Scientific American*, published on 17 November 1877. By 29 November Edison's mechanic, John Kruesi, had constructed a tinfoil phonograph, and on 7 December it was demonstrated to the editor of *Scientific American*, who then reported on it at length and gave a detailed description.[7]

The British public first learned about the phonograph from a report in *The Times* of 17 January 1878 based on information provided by Henry Edmunds, a British engineer who had been on a study tour in the USA and had seen Edison's phonograph. After reflecting on the wonders of modern science that had brought first the telephone and now the phonograph, it gave a detailed mechanical description which was read by William Preece, the Chief Engineer of the Post Office. He was giving a Friday Evening Discourse at the Royal Institution on 1 February 1878, and he arranged for the instrument-maker Augustus Stroh to make a phonograph from the description. He demonstrated it at the end of his lecture. On 27 February Preece gave a demonstration to the Society of Telegraph Engineers (later renamed the Institution of Electrical Engineers). For that he had three phonographs. One was a copy of Edison's original made by a Mr W. Pidgeon. One was Stroh's machine, which was driven by a falling weight. One was an improved model brought from America by Mr Puskas, a friend of Preece's.

We think of the gramophone as a medium of entertainment, but that thought did not arise with the phonograph in the 1880s. For some, it was a scientific instrument. In *Sensations of Tone*, Helmholtz had considered the nature of the vowel sounds. There were two rival theories, and in 1894 Lord Rayleigh pointed out that the phonograph might settle the matter. The method proposed was to record a vowel sound and reproduce it at different speeds to see whether the character as well as the pitch of the sound depended on the speed of rotation. The results were inconclusive, but the experiment helped to give the phonograph an academic respectability.[8] Edison saw the instrument as an adjunct to the telephone, or as a piece of office equipment. It could permit people without a telephone to record a spoken message and have it sent from a central telephone office in a manner analogous to telegrams. However, the quality of reproduction obtainable with a tinfoil phonograph was very poor and Edison soon lost interest in it.

It is no surprise that Alexander Graham Bell took a keen interest in the phonograph, and in 1880 he had the opportunity of working on the subject. He had been awarded the Prix Volta by the French government for his invention of the telephone. This prize had been established by Napoleon III, and with the money Bell set up the Volta Laboratory Association to carry out experiments in electro-acoustics. He invited Charles Sumner Tainter and his cousin Chichester Bell to join him. The result of their work was the graphophone, which was announced in 1885, after they had tried out many different recording arrangements, including magnetic recording. The principle of the graphophone was similar to that of the phonograph, but the sound impressions were cut in a wax coating on a cardboard cylinder. The cylinder could be removed but, unlike the tinfoil, it could also be replaced and played again.[9]

Edison's response to the new competition was a renewed interest in the phonograph, and the result was his 'perfected' phonograph, which appeared in June 1888 and which also recorded on wax. At the British Association in Bath on 6 September 1888, both an Edison perfected phonograph and a Bell graphophone were demonstrated. Colonel Gourard demonstrated the phonograph, and his visit to the British Association was part of a tour of Britain promoting the instrument. The graphophone was demonstrated by Henry Edmunds.

The essential principles of the two machines were identical: both incised the sound impressions in wax. In the graphophone the wax took the form of a thin layer on a relatively narrow cardboard cylinder, this record being used once only. In the phonograph the cylinder was wholly of wax and a thick version, not intended for posting, could be used many times over, since the instrument incorporated a shaver to erase a previously recorded impression. As regards the drive, the instruments were quite different. The graphophone used a treadle mechanism, with an ingenious governor to keep the speed of rotation reasonably constant. The phonograph used an electric motor, supplied from a three-pint bichromate cell, but most subsequent instruments used clockwork.

Neither the phonograph nor the graphophone was successfully exploited commercially for some years. Then they were soon superseded by the disc gramophone, and that only became a matter of serious electrical engineering interest in the 1920s when electrical recording was introduced.

ELECTRIC LIGHTING AND ELECTRICITY SUPPLY

From the late 1870s electric lighting was Edison's dominant interest, and to exploit the electric light he had also to develop an electricity supply. He sought to promote it in Europe, as well as in the USA, and especially in Britain where, as we have seen, he was already active. His British interests were managed by Edward H. Johnson, who was responsible for the Holborn Viaduct generating station in central London, which began supply in January 1882. This was Edison's first 'central' power-station, preceding his establishment at Pearl Street, New York, and was only the second public electricity supply undertaking in the world.

The year 1882 was a crucial one in electrical engineering history. Public supply had begun, the previous autumn, in the small English town of Godalming, though only a few householders took advantage of the new opportunity. Holborn Viaduct power-station began supply in January, and others were being planned. Legislation was being formulated to regulate the new industry, and the possibility of using electricity to transmit power as well as provide light was being studied.[10]

Edward Johnson, as Edison's representative in London, had to convince the financial and scientific establishments that Edison's system was the best and was going to dominate the new electric lighting market, and he had to convince Edison that he was doing all that could be done to promote Edison's interests. He felt he had scored a 'coup' when, in November 1881, he enlisted the services of John Hopkinson.[11] Hopkinson's name was already known to Edison, for when Edison was beginning his work on lighting he gathered the latest technical literature, including the work of Hopkinson.[12] Throughout 1882 the Edison interests had the benefit of Hopkinson's advice on patent matters, electrical machine design, distribution systems and the legislation then going through Parliament on electricity supply.[13] Their relationship was short-lived. Most of the surviving papers date from 1882, although the story is continued into 1883 in a memoir by Hopkinson's son Bertram, who had access to papers which no longer survive.[14]

John Hopkinson—two years younger than Edison—was born in 1849 in Manchester, where his father was a partner in an engineering firm. Educated at Owens College, Manchester and Trinity College, Cambridge, John Hopkinson first joined his father's firm and then became engineer and manager of the lighthouse department of Chance's optical works in Birmingham. In 1877 he moved to London to set up in practice as a consulting engineer, and soon established a reputation in the new but rapidly growing field of electricity supply and electric lighting. He acted as an expert witness in several legal cases, especially in patent matters. He was one of the judges at the first International Electrical Exhibition, held in Paris from August to November 1881.

Edison had developed a complete lighting system, including everything from the generator to the lamp, but he had serious rivals as far as the lamp was concerned, although he appears to have thought that he had a monopoly of the electric filament lamp. Joseph Swan, the Newcastle chemist, had also developed a viable filament lamp and he had an arguable claim to priority of invention. St George Lane Fox, an English aristocrat, and Hiram Maxim, the American better known for his work on firearms, were also making lamps. All used different filament materials. All four exhibited at the Electrical Exhibition in Paris. Edison received the highest honours from the exhibition jury, though there is evidence that his publicity activities contributed to that decision.[15] The numerical results published when the efficiency of the lamps was tested at the Exhibition suggested that although Edison's showed the highest efficiency there was not much to choose between the different makers.[16]

In late 1881 and early 1882 Edison and his English team had lengthy correspondence about patent matters. They were concerned about both the validity of some of Edison's English patents and the validity of some other people's patents on electric generator design. The English team sought the views of several leading patents barristers, and of John Hopkinson, whose views were all reported back to Edison.[17]

The most important case was *Edison* v. *Swan*. That case never went to a full trial, so the arguments were never made public. The essence of the case, however, was that Edison had a British patent for a filament lamp granted in 1879, and he claimed that Swan infringed it. Swan claimed that the patent was invalid, because he was already making lamps. In June 1882 the patents barrister Frederick Bramwell submitted a thirty-page report on the lamp patent.[18] Bramwell gave an exhaustive analysis of Edison's patent and concluded that it was valid and enforceable, but he pointed out a weakness where an opponent might attack the patent. Swan was the potential opponent.

An essential requirement of British patent law is that the patentee must give in his specification a description which is sufficient to permit a competent workman to carry out the invention without having to exercise inventive ingenuity himself. Edison had omitted a vital step from the description of the lamp manufacturing process in his specification, and the patent was therefore open to attack on the grounds that the specification was insufficient. That vital step was the process known as 'running on the pumps', in which a lamp is heated, by passing a current through it, while the evacuating process is being carried out. If the lamp is evacuated cold then, when it is first used, occluded gas is released from the filament and the vacuum is reduced. The discovery was an important stage in lamp development, and the argument would have been that it was not an obvious thing to do, given the state of knowledge in 1879, and therefore the specification ought to have mentioned it explicitly.

The 1883 case of *Edison* v. *Swan* did not get further than a preliminary action in which the High Court refused to grant an interim injunction against Swan. The Edison and Swan interests merged to form the Edison & Swan United Electric Light Company. Several years later the Edison & Swan United Company sued others for infringement of the same patent. It was then, of course, in Swan's interests to defend the patent which previously he had opposed. The defendants tried to get a court order to reveal the grounds on which Swan would have contested the patent in 1883, but they failed. It would be interesting to know if Swan's case against Edison was based on the weakness that Bramwell had pointed out.[19]

In a letter of 24 February 1882 Hopkinson described some of his own patents relating to electric motor control, and was clearly hoping to sell the American rights to Edison (he had already sold the British rights to Siemens).[20] Two of the patents were for motor control systems. The third was a combination of the control systems with a Weston friction clutch.[21] This is a mechanical arrangement used in lifts, hoists, etc. which holds the mechanism fixed when the motor is not driving, so that the load cannot run away. Edison obtained a report on the patents from his American patent attorney, Richard A. Dyer. In a letter dated 28 April 1882, Dyer said that some parts of Hopkinson's inventions (but only some parts) would be patentable in the USA. He also drew attention to some of the difficulties that American patent law placed on inventors who were foreigners living abroad, as compared with US citizens.[22]

Edison expressed interest in acquiring part of Hopkinson's patent rights. Writing to Johnson on 2 May he said:

> Referring to your favour of 26th February (which has remained unanswered first owing to my absence in Florida and second in consequence of my sickness since my return) I think the only thing in Dr Hopkinson's patents which would hold in this country and be of value is the Broad Idea . . . set in the third claim of that patent. . . . A Company of mine on Electric Rail Roads is about to be formed and I would advise Dr Hopkinson to take out a patent on the above referred to feature in this country with a view to making an arrangement with our Rail Road Coy which can readily be done if they decide to make use of the principle.

In subsequent correspondence Hopkinson offered the American rights to Edison for £3,000, though it is not clear whether this offer was taken up.[23]

There was an electric hoist, presumably using the patented ideas, at Holborn Viaduct. Several reports written during May and June mention the hoist, and Hopkinson thought it important to have an electrically powered machine visibly at work. The machine having been found to be working satisfactorily, it was handed over to J.A. Fleming[24] who was employed full-time by the Edison Company.[25] The generators used at Holborn Viaduct were made in the USA. Hopkinson was soon convinced he could improve on Edison's design. He made laboratory tests on small-scale models of the Edison machine and also on variant designs of his own. The induction in the armature in various designs was then compared by jerking the armature out of the magnet and observing the kick on a ballistic galvanometer connected to the winding. Hopkinson showed that in some of the Edison machines the armature and the field magnets were not magnetically saturated simultaneously, that a significant part of the effective area of the armature was lost by bolt holes and the like, and that shorter magnets of the same cross-section could give as large an induction with the same magnetizing current. Hopkinson urged the Edison Company to have a machine made with shorter magnets and an armature of greater effective cross-section. The resultant machine had an output more than twice that of the Edison machine of the same weight and cost, from which it was evolved. Known as the Edison–Hopkinson machine, it was manufactured by Mather & Platt, and was the pattern which other manufacturers followed.[26]

EDISON IN THE BRITISH PRESS

The Holborn Viaduct generating station was closed down in 1884, so when Edison made his second visit to Britain none of his plant was in operation. His visit was followed closely by the press, and the *Daily News* ran a special series of articles on electric light in London.[27] Edison was quoted frequently. Clearly he represented electricity as far as the press was concerned, and his views and thoughts were of interest. 'Mr. Edison's dream is to light up the New World at night with the brightness of day, by electricity from the Atlantic waves. Every time he crosses the ocean he turns sick at the thought of all that force running to waste.'

Edison made two visits to London electricity undertakings, the Metropolitan Electric Supply Company, which owned six stations in the West End, and the London Electric Supply Corporation, whose massive generating station, then being built at Deptford, was the brainchild of Sebastian Ziani de Ferranti, whom the *Daily News* dubbed 'the Edison of England'.[28] His first visit was to three of the stations of the Metropolitan company, though illness nearly prevented it.

> This enterprising company has been opening its stations in good time for a flying visit from Mr. Edison, who is passing through London. Owing to Mr. Edison's really severe indisposition, it was doubtful, even up to yesterday morning, whether the renowned inventor and discoverer would be able to fulfil his intention of inspecting the 'Metropolitan's' centres. But he appeared early and quite unexpectedly. It would be difficult, at first sight, to detect signs of illness in the firm, erect, shortish figure. The fresh, beardless, almost boyish face, however, bore traces of ailment. His sciatica was well nigh gone, though he was suffering from a bad cold. Still, he was in a capital humour. The only aged thing about him is the colour of his hair—which the forty-two winters of his life have bleached pretty well. Simple, frank, unaffected, thoroughly good natured, and ever alert is Mr. T.A. Edison, of both hemispheres. Mr. Edison inspected only three of the 'Metropolitan' Company's stations: namely, the station in Rathbone-place, the station in Sardinia-street (behind Drury-lane), and the station in Whitehall Court. Besides these three, there is a station each at Waterloo Wharf, Manchester Square, and Greenmore Wharf. But the three first named are the only ones yet ready.

The Sardinia Street Station had gone into commission less than three weeks earlier, on 1 September 1889, and it was not to last very long. On 24 June 1895 it was practically destroyed by a fire which broke out behind the switchboard, and of which the cause was never satisfactorily determined. It would have been of special interest for Edison, however, since it was the one station then working in London which had American equipment, and Edison was reported to have 'expressed himself thoroughly well pleased'. The machinery was obtained from Westinghouse, of Pittsburgh, and consisted of ten belt-driven single-phase alternators each giving 125 kW at 1,000 V. Before the fire the company had already ordered four 350 kW Parsons turbo-alternators to replace some of them, and more were quickly ordered.[29]

It was reported that Edison did not agree with British engineers about distribution voltages:

> Without entering into technicalities, we may say that the generators of domestic electric supply in America proceed upon the low pressure system. The 'Metropolitan' Company and the 'London' Supply Company proceed on the high pressure system. The London men say that what may suit Mr. Edison and his friends in New York does not suit customers and manufacturers (if that is the correct designation) here. The low pressure system requires a large number of supply centres. A little parish like St James's, for instance, would require three. But a multiplication of stations would be costly; the rents would be enormous. With a low pressure system in London the electric circuit would extend no longer than from eight hundred to a thousand yards. In New York there is comparatively little trouble in getting ground for premises and cables, even if rents be high. In London it is very different. Now, with a high pressure system, an electric company may send its circuits over miles and miles, without any loss of force. With its immense 'tension', the London Company can already send a current through seventy-three miles of cable, and with absolute safety to all concerned. For this high pressure system a single large central station is, obviously, the best source of distribution. The 'Metropolitan' Company has adopted a medium course. Instead of generating from one central station, such as the 'London's' at Deptford, it will supply, as already said, from half-a-dozen. The 'Metropolitan' has also refrained from committing itself to any single type of machinery.

Edison's ideas were even further removed from those of Sebastian Ziani de Ferranti, who was then building his massive generating station at Deptford and hoping to supply a large area of London through 10,000-V mains from his single station. On Wednesday 25 September Edison visited the construction site, and it was quite an occasion:

> In front of the works there was a gay and liberal display of bunting, in honour of the distinguished visitor. Among those who accompanied him from town, or who were on the spot to meet him, were Sir John Pender, with whom Mr. Edison has been staying during his short visit; Mr. Forbes; Mr. Ferranti, the London Company's engineer, whom we have dubbed the Edison of England; Colonel Gouraud,

> Mr. Verity, Mr. Pender, Dr. Hopkinson, Mr. Gordon, Mr. Ince, Mr. Pyke; Mr. Claremont, of the 'Metropolitan' Company; Mr. Cunliffe Owen; Mr. Bain; Sir Coutts Lindsay, of whose place in the history of electric lighting we have formerly spoken; and the present writer. It was a delightful day—for a sharp walk; but the wind from the river 'bit shrewdly.' Mr. Edison felt it, for he was still suffering from his severe cold, he looked a dozen years older when, escorted by Mr. Ferranti and Sir John Pender, he entered the gateway, with his coat collar up to his ears, white muffler about his neck, and his hands stuck in his pockets. His face was much paler than usual. But he soon appeared to forget his indisposition when he began his inspection of the wonderful machinery. He spent a long time minutely examining the immense engine and dynamo which, it is hoped, will in a few days will begin to supply lighting current over one section of the London region which has been assigned to the company. But as to the outfit of the works, as they will be several months hence, when the ten thousand horsepower engines are erected in the lower area of the building, Mr. Edison could only infer what it will be from the plans on the spot and from Mr. Ferranti's description. The diameter of each of the huge dynamos yet to be erected will be no less than forty-two feet! and each dynamo will be composed in a circle of 192 enormous magnets. But to most people the Deptford works would at present be little better than a chaos. Some study of them would be required to enable one to realise what they would be like when in complete order. Mr. Edison went over the whole ground, to the river front of the works. In spite of the demoralization of cold, Mr. Edison admired the view from this spot—the ocean ships and steamers moored in the river, Greenwich Hospital to the left, and in front—the Isle of Dogs. A long way behind the Isle of Dogs, and nearly opposite the hospital, loomed the unadorned Brobdingnagian form of the biggest gasometer in the world, and as far as one knows in the Solar System. The gasometer and Mr. Edison are symbols respectively of the old and new orders of things illuminant. But though the Edisons and the Ferrantis are taking the shine out of the gasometers, they are not necessarily taking it out of the gasometer shareholders' pockets. That big gasometer, I was told on the spot, is about to increase its premises. There are many other uses for gas besides lighting, and the gas companies are finding them out.

On the way back to London Edison was asked what he thought about electrical progress in Britain. 'You may be slow to begin,' he replied, 'but I must say that when you do go ahead you may even beat us. The next thing you will have to do will be to turn your electric force into motive power. America is the country for that; in New York and other places, even small shopkeepers use electricity in scores of ways, in place of hand labour and old fashioned mechanical appliances; they can buy it cheap—for a few cents for that matter. Electricity is not used as it should be, for lifts as an instance, in this country.' The discussion then turned to electric traction, where Edison thought Britain was far behind. He talked about the electric railways being developed in America, but was apparently unaware of the City and South London Railway then under construction.

The final comments were on the importance of giving customers a reliable supply:

> A gentleman of the party was remarking that if 'once' a breakdown happened, public confidence in the new light would vanish. 'Twice,' said Mr. Edison, on the instant, with his quiet smile; 'the first time the public will excuse you, they will make allowance for inexperience, but if you do it a second time you are done for.

Acknowledgements

I am grateful to the team of the Edison Papers Project, Rutgers University, New Brunswick, New Jersey, especially Reese Jenkins, Paul Israel, Keith Nier, Melodie Andrews and James Spiller, for guidance through the Edison Archives and many helpful discussions; to Professor James Greig, formerly professor of electrical engineering at King's College London, for discussions about Hopkinson; and to Robert Harding for guidance with the Hammer Collection. Tom Hughes, Frank James and Lenore Symons have discussed aspects of this work also, and I am grateful for their cooperation.

Notes and References

1. Reese V. Jenkins *et al.* (eds), *The Papers of Thomas A. Edison*, vol. 1, (Baltimore and London, 1989), 224. (Cited hereafter as *Edison Papers.*)
2. *Edison Papers*, 250–2.
3. *Edison Papers*, 514–15.
4. Summary: *Edison Papers*, 591–9; full details: Post Office Archives, 'Automatic Telegraph' file.
5. *Edison Papers*, 36, and 'Headnote', 30.
6. *Edison Papers*, vol. 1, 659, 36n. 1.
7. The visit is reported in *Scientific American*, 22 December 1877, 384–5.
8. V.K. Chew, *Talking Machines*, (London, 1973), 8.
9. For Bell generally see Robert V. Bruce, *Alexander Graham Bell and the Conquest of Solitude*, 1973.
10. For fuller details see Brian Bowers, *A History of Electric Light and Power*, especially chapters 9 and 10.
11. Letter, Johnson to Edison, 3 November 1881. The letter is reproduced in *The Thomas A. Edison Papers, Selective Microfilm Edition*, (University Publications of America, Frederick, Maryland, USA, Part 1, 1985, Part 2, 1987), reel 58, frames 682–9. References to the microfilm edition are cited hereafter as TAEM.
12. There are notes from Hopkinson's published work in Edison's notebooks. See, for example, TAEM, 32: 265, 270 and 273.
13. The relationship between Edison and Hopkinson, and Hopkinson's role as consultant to Edison on electricity distribution systems and in connection with the pending legislation, is the subject of another paper: 'Edison and Hopkinson—transatlantic relations in electrical engineering in the early 1880s', presented at the Second International Conference on the History of Electricity, Paris, July 1990, and to be published in the *Proceedings* of that conference by the Association pour l'Histoire de l'Électricité en France.

14. Bertram Hopkinson (ed.), *Original Papers by the Late John Hopkinson DSc FRS*, two vols, (Cambridge, 1901). The *Memoir* is at the front of the first volume.

15. R. Fox: 'Edison et la presse française lors de l'Exposition d'électricité de 1881', *Un Siècle d'Électricité dans le Monde*, (Paris, 1986), 223–35.

16. Bowers, op. cit. (10), 120.

17. See letter, E.H. Johnson to Edison, 16 January 1882, in the Hammer Collection in the National Museum of American History, Washington, DC, especially page 5.

18. Report by Frederick Bramwell. TAEM, 62: 955–84.

19. For a study of the legal action between Swan and Edison, and the surrounding circumstances, see C.N. Brown, *J.W. Swan and the Invention of the Incandescent Filament Lamp*, (London, 1978).

20. Letter, Hopkinson to Johnson, 24 February 1882, TAEM, 63: 400.

21. Memorandum by Hopkinson, dated 24 February 1882, on patents for the Transmission of Power by Electricity. TAEM, 63: 401–5. The text of the patent specifications is in TAEM, 63: 406–19. The three British patents by Hopkinson were: No. 2481/21 of June 1879 for reversing the direction of a motor by changing the points of contact of the brushes. No. 4653/14 of November 1879 'for alternative methods of reversing the direction of rotation viz (1) using the two opposite faces of a single brush as an equivalent of the double brush of the first patent and (2) sliding the brushes round the axis of the armature with or without breaking the armature circuit by the movement. The method of the former plan is probably the best as the commutator drum runs from the point of support of the brush and the brush is therefore less likely to trip, it was used by Siemens at the Paris Exhibition both on their tram car and on the hoist which was located in the corner near the Brush machines.' (There was a corresponding American application which had been refused, but which Hopkinson thought might be worth appealing.) No. 2989/7 of July 1881 for 'Transmission of power', which describes a combination of the motor control systems above with a Weston friction coupling. No corresponding American application had yet been made. Hopkinson ends: 'Messrs Siemens Bros have an exclusive licence under the three English patents, the terms including a liberal minimum royalty. As to my rights in America arising from the three English patents my wish is to transfer them for a suitable price to someone who can use them more efficiently than I can.'

22. Letter, Dyer to Edison, 28 April 1882. TAEM, 63: 451–4. The legal point here is that under American law an inventor living in the USA can get a patent for something if he can prove that he had the idea before anyone else. For an inventor living abroad the only acceptable evidence of date was a foreign patent application. For this reason many of Edison's notes and drawings are signed and dated by witnesses, in case the ideas they contained were to become the subject of patent applications.

23. Hopkinson's basis for the price asked was: 'Messrs Siemens Bros have an exclusive licence in this country for all three patents paying 5% royalty on all machines to which the device may be applied and a minimum of £300 per annum without power of revocation unless the patents are upset. What I would suggest is that Mr Edison should take up all my rights in America based upon these three patents whatever they may be on the following terms: Mr Edison to apply at once for the two patents suggested by Mr Dyer based on my 1881 patent and if he think fit for patents in interference with Field. Mr Edison to elect on or before 1st October of this year whether he will take the patents applied for or not. If he elects to take them he should pay me £3000 for their complete purchase, if he declines to take them he should assign the patents obtained back again to

me free of any cost to me. I fix the sum of £3000 on the ground that the present value to me of the Siemens licence is at least £6000 and I halve that sum in consideration of my difficulty as to the American patents.'

24. Report by Hopkinson dated 7 June 1882, section 1. TAEM, 62: 950–4.

25. Born in 1849, and educated at University College London, he worked with James Clerk Maxwell at the Cavendish Laboratory, Cambridge. He became Professor of Physics and Mathematics at University College Nottingham in 1881, but resigned after a year to join the Edison Company. He did not stay long, however, and in 1885 became Professor of Electrical Technology at University College London, a post he held for forty-one years. He worked on telegraphy, and with Marconi on radio. In 1904 he experimented with the Edison effect, and made a new device which he called the 'thermionic valve'. He died in 1945 at the age of ninety-five. See Charles Susskind: 'Fleming, John Ambrose', Charles C. Gillespie (ed.), *Dictionary of Scientific Biography*, (New York, 1972).

26. Bertram Hopkinson, op. cit. (14), xliii. He cites a report dated 17 May 1883 from Hopkinson which appears not to be in the Edison archives.

27. The articles appeared on Friday 20 September, Monday 23 September, Wednesday 25 September, Thursday 26 September, Monday 7 October and Saturday 12 October 1889, and the press quotations in this section are all from these.

28. *Daily News*, 23 September 1889, 3.

29. R.H. Parsons, *The Early Days of the Power Station Industry*, (1939) 75–8.

The Contributions of the Bell Telephone Laboratories to the Early Development of Television

R.W. BURNS

INTRODUCTION

Television was first privately demonstrated, in a very rudimentary way, by J.L. Baird on 2 October 1925 and subsequently was shown by him to approximately forty members of the Royal Institution on 26 January 1926.

Baird's achievement, the accomplishment of which had occupied his attention from the winter of 1922–3,[1] was the realization of a quest which had engaged the minds of a considerable number of scientists, inventors and engineers from 1873. His work commenced at an opportune time, for by the early 1920s all the basic components of a crude television system appeared to be available. Whereas before 1920 only a few isolated attempts had been made to investigate, on an experimental basis, the subject of 'distant vision', from about the beginning of the 1920–30 period determined efforts to advance television were being effected in the UK, the USA, France and Germany. Initially these endeavours were mainly those of individuals working in isolation from others. J.L. Baird of the UK, C.F. Jenkins of the USA and D. von Mihaly, a Hungarian working in Germany, were three of the principal early investigators in this period. For several months in 1923 Zworykin pursued some personal work on an all-electronic television camera while he was at Westinghouse Electric and Manufacturing Company, USA, but the only determined effort being made in that year by a public or private organization seems to have been that which was originated at the Admiralty Research Laboratory, Teddington, Great Britain.[2]

From 1925 this situation changed. Bell Telephone Laboratories, of the American Telephone & Telegraph Company (AT&T), began an ambitious programme of work which led to an impressive demonstration, in April 1927, of well engineered apparatus for the transmission and reception of television images by land line and radio links. From 1925, General Electric, Westinghouse Electric & Manufacturing Company and the Radio Corporation of America, in addition to several smaller companies, began to be associated with television projects; in Germany both Fernseh AG and Telefunken were active in this field by the end of the decade. Leading companies in the UK adopted, until 1930, a rather reserved position on television matters and before then only the Baird companies (Television Ltd, Baird Television Development Company and Baird International Television Ltd) vigorously engaged in the pursuit of 'distant vision' research and development. The Marconi Wireless Telegraph Company and the Gramophone Company (later part of EMI Ltd) commenced their television activities in 1930.[3]

The results that were obtained by the Bell Telephone Laboratories in 1927 are particularly important historically because they represent the best that could be expected with the technology as it existed at that time. With their vast resources in finance and equipment, and in staff expertise and experience, the laboratories were uniquely able to demonstrate what could be engineered in the field of television. Subsequently colour television and two-way television systems were realized in 1929 and 1930 respectively, all at great cost. From 1925 to 1930 (inclusive) AT&T approved the expenditure of $308,100 on low definition television.

After 1930 the Bell Laboratories continued their work on television but without achieving successes of the type which were contemporaneously being manifested by RCA of the USA and EMI of the UK, despite the allocation of $592,400 to the work from 1931 to 1935 (inclusive). Thereafter television research and development declined and ceased to be part of the Bell Laboratories' interests sometime in 1940.

This paper considers the contributions of Bell Telephone Laboratories to television progress.

HISTORICAL BACKGROUND

This section is based on one of my earlier papers.[4] In 1873 Willoughby Smith communicated a discovery to the Society of Telegraph Engineers which appeared to provide, a few years later, a means for 'seeing by electricity'. Smith's disclosure concerned the photoconductive property of selenium, a characteristic of the element which was utilized in many of the early schemes for television until the development of suitable amplifiers and photoemissive cells made selenium cells obsolete in the 1920s.

Before 1873, several methods had been advanced for transmitting images of line drawings and printed sheets, commencing with the inventions of Alexander Bain in 1843 and Frederick Bakewell in 1848 and culminating in the operational picture telegraph schemes of Caselli (1862), Meyer (1869) and d'Arlincourt (1872). But, with all these systems, the lack of a suitable photoelectric transducer had limited their application to the reproduction, at a distance, of images of pictures and the like which could be drawn in varnish on a conducting sheet or which could be suitably prepared so that a scanning stylus and associated apparatus could discriminate between the picture and the background.

During the period 1843–72, the only known effect which related changes

Table 1 Dates (and names of inventors) of some 'distant vision' proposals for the period 1878–1924

1878	de Paiva	1904	von Jaworsky and Frankenstein	1920	Kakourine
1879	Perosino			1920	Egerton
1879	Senlecq	1904	Ribbe	1921	Whiston
1880	Carey	1906	Lux	1921	Schoultz
1880	Ayrton and Perry	1906	Rignoux	1922	Belin
1880	Middleton	1906	Diecjmann and Glage	1922	Valensi
1880	Sawyer	1907	Rosing	1922	Jenkins
1880	Le Blanc	1908	Campbell Swinton	1922	Rtcheouloff
1881	Senlecq	1908	Adamian	1923	Baird
1881	Bidwell	1908	Anderson and Anderson	1923	Hammond
1882	Lucas	1908	Sellers	1923	Zworykin
1884	Nipkow	1909	Ruhmer	1923	Gardner and Hineline
1889	Weiller	1910	Schmierer	1923	Nisco
1890	Sutton	1910	Ekstrom	1923	Western Electric
1893	Pontois	1910	Hoglund	1923	Stephenson and Walton
1894	Majorana	1911	Rosing	1923	Robb and Martin
1894	Jenkins	1911	Campbell Swinton	1924	Sequin and Sequin
1895	Nystrom	1911	Rosing	1924	Blake and Spooner
1897	Szczepanik	1914	Lavington Hart	1924	Alexanderson
1898	Vol'fke	1915	Voulgre	1924	Hoxie
1898	Dussaud	1915	Dauvillier	1924	Takayanagi
1899	Bolumordvinov	1917	Nicolson	1924	Apollinar and Zeitlin
1902	Coblyn	1919	von Mihaly	1924	McCreary
1902	von Bronk	1919	Sandell	1924	d'Albe
1903	Belin and Belin	1920	Baden Powell	1924	Dieckmann

of light intensity to changes in electric current was that discovered by Becquerel in 1839 in his investigations on the electrochemical effects of light. Becquerel's observations demanded the use of a highly sensitive galvanometer and consequently the effect was not appropriate for incorporation into a 'distant vision' scheme.

The photoconductive property of selenium was easily demonstrated ('its sensibility to light is extraordinary, that of a mere lucifer match being sufficient to effect its conductive powers'), and in the decade following Willoughby Smith's letter there was an expectation among scientists and others that 'seeing by electricity' would soon be a reality. This expectation was based not only on the results that had been achieved in the field of picture telegraphy, together with Smith's disclosure, but also on the invention of the telephone, by Alexander Graham Bell in 1876, which enabled 'hearing by electricity' to be readily implemented. The simplicity of Bell's device and the lack of extended effort involved in its development possibly stimulated inventors to attempt the transmission of moving images by electrical means, for numerous suggestions for 'telectroscopes' were put forward in the fifty-year period following Smith's important announcement (see Table 1).

The problem of 'seeing by electricity' was, however, of an altogether different order of complexity compared with the problem of 'hearing by electricity', and success eluded the early workers. An indication that selenium was not an ideal photoconductive material was given by Sale in a communication to the Royal Society in 1873. Sale's experiments indicated that instantaneous changes of light intensity on a bar of selenium did not cause instantaneous changes of resistance in the material.

Nevertheless, the work of the nineteenth-century television pioneers was not wholly unproductive and, by the end of the century, some of the basic system components needed to engineer a television plan had been proposed. The elementary principles of scanning in particular were well understood and the scanners of Nipkow (1884), Weiller (1889) and Brillouin (1891) were later successfully utilized by many scientists and engineers in the period 1925–36. The development effort which had enabled practical picture telegraph systems to be demonstrated and introduced into the public service (albeit for short periods in the 1860s and just before and after 1900) had given inventors an understanding of the principles and difficulties of synchronization. This understanding was applied to the problem of television; indeed the use of line synchronizing pulses in modern television can easily be traced back to the work of Bain in 1843.

After the turn of the century the notions of Rosing (1907) and Campbell Swinton (1911) on the employment of cathode ray tubes in distant-vision schemes provided the new ideas which were needed to achieve an all-electronic solution to the television problem. In addition, Hallwach's demonstration of the photoelectric effect (1888), the detailed investigations of Elster and Geitel (1889–1913) on photoelectricity and de Forest's invention of the audion (1907) were salient contributions that were to play a vital part in the progress of television.

De Forest's invention, the triode valve, undoubtedly hastened the time when seeing by electricity became a reality. In 1912 de Forest discovered that the valve could be employed in an oscillator to generate electromagnetic waves in addition to acting as a detector and as an amplifier. This finding was to prove of great significance in the history of sound and vision broadcasting. The Marconi spark apparatus, the Poulsen arc and the Alexanderson alternator were all expensive and cumbersome. But with de Forest's valve generator and the stimulus for invention and innovation provided by the First World War, the progress of continuous wave radio communications rapidly moved forward to the stage where commercial sound broadcasting could be seriously contemplated shortly after the cessation of hostilities in 1918. And if sound signals could be propagated by radio then surely vision signals too could be transmitted.

The first broadcasters were the radio amateurs or 'hams', whose hobby it was to communicate with other hams. Their activities were curtailed in the UK during the 1914–18 period to prevent interference with essential communications, but in 1919 (in the UK) the restrictions were lifted. In the USA the enterprises of the amateurs were not curbed by the war and hence the establishment of regular broadcasting could proceed more quickly than in Great Britain.

By the end of 1920 sound broadcasts had been transmitted from stations in the USA, the UK, France, The Netherlands and other Western European centres of population. Essentially the birth of the new form of entertainment took place in 1920 because the conditions necessary for its success were opportune at that time. The growth of commercial radio telephony and domestic broadcasting influenced the progress of television. Whereas only a few new schemes for seeing by electricity were put forward (for example, by Lavington-Hart, Voulgre and Nicolson) during the 1911–20 period, in the next decade television development was to advance rapidly and become a reality.

BELL TELEPHONE LABORATORIES

Although the Bell Telephone Laboratories initiated a television project in the mid-1920s there is some evidence that the Bell System had an interest in television during the First World War and in 1921.

In November 1912, A.McL. Nicolson,[5] a British subject living in New York, joined the Research Branch of the Engineering Department of the Western Electric Company (now Bell Telephone Laboratories) and soon after began active work on vacuum valve research in association with Dr H.D. Arnold and Dr H.J. van der Bijl. Nicolson's first important contribution was the development in 1913 of the 'unipotential' (or indirectly heated) cathode, an innovation that achieved worldwide usage. His coated-filament manufacturing process was employed in the fabrication of all telephone and radio valves made by the company before 1917.

Later, from 1917, he undertook a major study of piezo-electricity and was the first person to discover that the mechanical vibrations of a crystal such as Rochelle salt could be coupled to an electric circuit by means of suitable electrodes, and that the resulting electromechanical vibrating system performed exactly like an electrically tuned circuit. He successfully used such crystals in loudspeakers, microphones, gramophone pick-ups and valve oscillators.

Nicolson was a prolific inventor throughout the early years of radio and television, and during his life applied for approximately 180 patents. On 13 October 1916 he forwarded to E.H. Colpitts, the Head of the Research Branch, a 28-page paper[6] on television transmission and reception and mentioned in his covering letter: 'This completes my work on the subject as first submitted to you July 23, 1915.' At the end of the paper Nicolson referred to original manuscript material of 95 pages (with 170 figures) dating from 20 April 1916 to 7 October 1916. Unfortunately his paper gives no indication of the stimulus which led to his study of television, but it does list eleven features of his system which were considered to be patentable. Subsequently, on 7 December 1917, Nicolson sought patent protection for his method of television and this was granted on 16 October 1923; he assigned the patent to the Western Electric Company of New York. The company evidently felt the patent had some worth for it obtained patent safeguards not only in the USA but also in France and the UK.[7] Nicolson's scheme included an electromechanical transmitter and cathode ray tube receiver and was one of the first to embody thermionic valve circuits operating as amplifiers and oscillators. Furthermore, the patent clearly illustrated wireless broadcasting and reception of the vision signals.

Seventy-three claims were lodged by Nicolson in his US patent and thirteen in the equivalent British patent. Of these claims the most important from the point of view of the development and history of television concerned his use of a grid, in the cathode ray tube, to control the intensity of the electron stream. Before Nicolson's 1917 patent application, inventors such as Dieckmann and Glage,[8] Rosing[9] and Campbell Swinton[10] had incorporated post-anode control of the intensity of the electron beam but now a new and much more effective means of control was suggested for this purpose. F. Skaupy[11] in 1919 also advocated the utilization of grid control in telephotographic systems, and the method was mentioned by P. Villard in his 1908 book, *Les rayons cathodiques*. Zworykin[12] put forward the same method in his US patent of 13 July 1925 and it is now a standard feature of all modern television receiving tubes.

Another aspect of Nicolson's patent which is of some significance is his proposal to use a single carrier wave for propagating both the picture and synchronizing signals. Previously the transmission link between the transmitter and receiver of notional television schemes had been a transmission line. However, with the rapid advances which were taking place in the field of communications, a new mode of wireless transmission was available. Zworykin's 1925 patent likewise included a wireless link, but as the Western Electric Company had filed a prior claim, Zworykin necessarily had to use a different transmitting arrangement. His solution was to employ two carrier signals, of different frequencies, one of which was modulated by the picture signals and the other by the synchronizing signals, both being radiated from a common antenna.

The remaining aspects of Nicolson's patent hardly advanced the state of the art, although his method of deriving positional signals by the use of a double photocell was ingenious, if of doubtful realization. Oscillating mirrors were never popular with large industrial organizations working in the field of television, and the simple, versatile, robust and cheap Nipkow disc, and the more costly mirror drum, were never to be supplanted by oscillating devices, which perforce tended to be costly, fragile and limited in their performance as scanners.

Apart from Nicolson's report and patent the only document that relates to early (pre-1920) television interest in a Bell company is the memorandum[13] written in January 1894 by G.K. Thompson of the Mechanical Department of the Boston Laboratory of the American Bell Telephone Company. Thompson advocated the employment of Nipkow disc scanning analysers and synthesizers, an electric shutter type of light valve at the receiver and a light-sensitive cell comprising a large number of glass bulbs containing burnt cork [*sic*] at the transmitter. He thought the subject was 'worthy of being considered by [the] Company'.

Following the birth of broadcasting, in about 1920, interest in television was renewed. On 6 September 1921, P.M. (probably Pierre Mertz) wrote an eight-page memorandum[14] 'to note down some preliminary considerations regarding a possible system of television'. Whether this was a personal

Table 2 The results of P.M.'s calculations on multi-channel picture transmission

No. of dots per inch (n)	*Total no. of dots ($25n^2 \equiv N$, say)*	*Signalling speed for one channel ($16N \equiv S$, say)*	*Multi-channel*	
			No. of channels (5n = no. of rows)	*Signalling speed per channel = S/5n*
150	562,000	9,000,000	750	12,000
100	250,000	4,000,000	500	8,000
60	90,000	1,450,000	300	4,800
16	6,400	102,400	80	1,280

initiative or was requested by the company was not stated. His paper was read and very briefly annotated, by a person having the initials A.W., on 1 July 1923. P.M. examined the prospect of transmitting picture signals by means of multi-channel communication methods. For a picture of 5 × 5 inches, and using half-tone printing screens capable of printing 150, 100, 60 and 16 dots per inch, P.M. calculated the number of 'dots' which would have to be sent every second (the signalling speed) for a televised image to be synthesized. Then, for a number of signal channels equal to the number of dots in a row of the picture, he determined the signalling speed per channel. His results are given in Table 2.

'These requirements are extremely severe,' noted P.M. His observation was similar to that obtained, in 1908, by Shelford Bidwell who, by a congruous method of calculation, found that 160,000 synchronized operations per second would be needed to transmit a 2 × 2 inch image. This was 'widely impracticable', he opined.[15] His view led to Campbell Swinton's all-electronic television notions of 1908 and 1911.

There appears to be no evidence to show that either P.M.'s or Nicolson's ideas were ever implemented. Rather, AT&T commenced its experimental study of the television problem when 'it began to be evident that scientific knowledge was advancing to the point where television was shortly to be within the realm of the possible'.[16] The company was of the opinion that television would have a real place in world-wide communications and that it would be closely associated with telephony. It was certainly well placed to advance television, not only because of the extensive facilities of the newly formed (1925) Bell Telephone Laboratories but also because of the experience acquired in the R&D work which had made trans-continental and transoceanic telephony and telephotography possible.

In January 1925 development work under the direction of Dr H.E. Ives had been completed on a system for sending images over telephone lines and so research resources and expertise existed for a new scientific venture. Dr Ives and Dr Arnold, the Director of Research, agreed that the next problem to be undertaken was television. 'At Arnold's request,' wrote Ives,[17] 'I prepared and submitted to him on 23rd January 1925, a memorandum surveying the problem and proposing a programme of research.'[18]

Ives was eminently well qualified to lead a television project team. His experience and erudition at that time (1925) had been founded on work and investigations on colour photography, phosphorescence, illumination, colour measurement, intermittent vision, photometry, photoengraving, photoelectricity and picture transmission. His standing in his chosen fields had been recognized by the award of three Longstreth Medals (in 1906, 1915 and 1918) by the Franklin Institute of Philadelphia. Later he was to receive three more medals: in 1927 the John Scott Medal, in 1937 the Frederick Ives Medal of the Optical Society of America, and, after the Second World War, the Medal for Merit, the highest civilian award of the US government, for his war work.[19]

Ives's memorandum discussed the characteristic difficulties of securing the requisite sensitiveness of the pick-up apparatus, the wide bandwidths which from his experience of picture transmission were indicated as necessary for television, the problem of producing enough modulated light in the received image to make it satisfactorily visible, and the problem of synchronizing apparatus at the sending and at the receiving ends of the transmission link. The memorandum concluded with a proposal for 'a very modest attack' on the problem, capable, however, of 'material expansion as new developments and inventions materialized'.

Ives felt that these difficulties could be examined by utilizing a mechanically linked transmitter and receiver, each incorporating a Nipkow disc scanner operating on a fifty lines per picture, fifteen pictures per second standard. A photographic transparency, later to be superseded by a motion picture film, would be used at the sending end, together with a photoelectric cell and a carbon arc lamp. At the receiving end Ives proposed the use of a crater-type gaseous glow lamp. His plan was thus based on the transmission of light through the 'object' rather than on the reflection of light from an opaque body: the latter problem was found by experimenters to be much more difficult to solve than the former problem. A sum of $15,000 was approved for the project.

As noted previously, the principal, active, television experimenters from 1923 were C.F. Jenkins of the USA, D. von Mihaly of Hungary and J.L. Baird of Great Britain.[20] The approaches of the three inventors were individualistic. Jenkins was a well-known inventor and a person of considerable means. He had produced important inventions in the field of cinematography and was able to design and manufacture equipment of some complexity. His rotary scanners consisted of either specially ground prismatic discs or costly lensed discs. Mihaly, an experienced patent expert and engineer, used an oscillating mirror scanner, together with tuning forks and phonic motors for synchronizing purposes. Baird's impecunious state, together with his lack of R&D experience in electrical engineering and his lack of laboratory and workshop facilities (he conducted his experiments in the unsuitable conditions of private lodgings), constrained him to implement his plans with the simplest scanner that had been devised, namely the Nipkow scanner. By the late 1920s Mihaly and Jenkins, and others, had adopted this type of scanner for their work. It endured for many years: even after the adoption of the all-electronic television system of

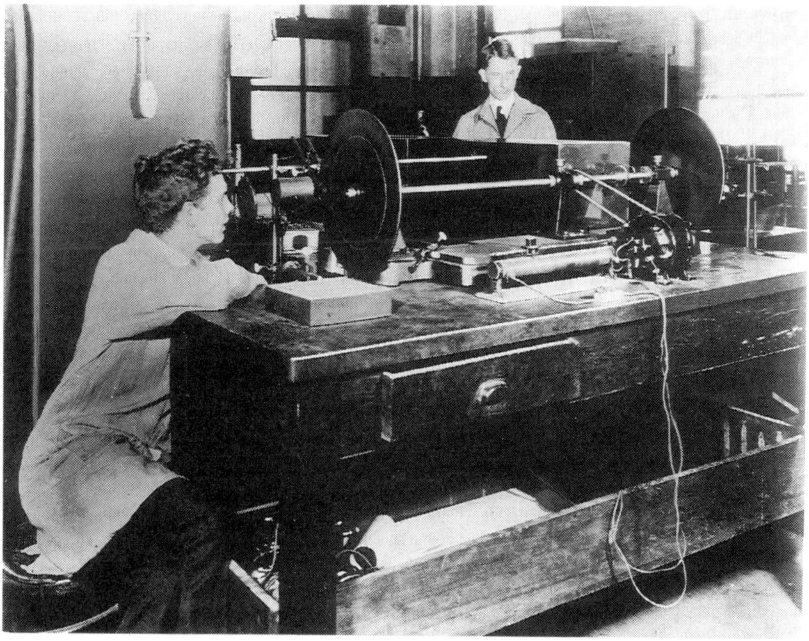

Figure 1 Research apparatus used in the development of television. Through a peephole J.R. Hefele is observing the image recreated through the rotating disc. The scanning disc at the other end of the shaft intervenes between an illuminated transparency and the photoelectric cell. The latter is in the box which is visible just beyond the driving shaft. Behind this box stands Dr Frank Gray.

the Marconi-EMI Company by the BBC in 1936, the Nipkow disc scanner was still being used in some television schemes.

By May 1925 the apparatus for Ives's design had been constructed and was in operation. A memorandum[21] of 14 May 1925, by J.G. Roberts, a patent attorney, records: 'I witnessed today a demonstration of Mr Ives's system of television. He has constructed and put into operation substantially the system he described in his memorandum of 23 January 1925, to Mr Arnold. In viewing the picture at the receiving end, I could distinguish with fair definition the features of a man's face like that of a picture at the transmitting end and also observed that, when the picture at the transmitting end was moved forward or backward, or up or down, the picture at the receiving end followed these motions exactly.' (See Figure 1.)

With this initial success behind them Ives's group, Dr F. Gray, J.R. Hefele, R.C. Mathes, R.V.L. Hartley *et al.*, next tackled the problem of synchronization when the two Nipkow discs were uncoupled. H.M. Stoller was given responsibility for this particular phase of the project and, by December 1925, the group was able to show motion pictures from a projector driven in synchronism with the discs. Another stage in the research programme was passed on 10 March 1926 when, at the conclusion of the ceremonies to mark the fiftieth anniversary of the invention of the telephone by Alexander Graham Bell, F.B. Jewett, President, and E.B. Craft, Executive Vice-President, talked over a telephone circuit in the telephone laboratory and were able to see each other's faces on screen at either end of the line.

According to Ives,[22] the group forbore to announce their achievement because, from the beginning of their investigations, it had been considered that only when vision signals could be sent over large distances—to parallel 'what had been done for voice signals'—would their apparatus be worthy of the appellation 'television system'. 'It would be television when the laboratory experiment was expanded to cover distances beyond any the eye could reach.'

By 7 April 1927 the system was ready to be demonstrated. It has been estimated that over one hundred engineers, scientists and technicians contributed to the success of the project,[23] though some reports mention a figure of one thousand. The demonstration,[24] using a wire link, consisted of the transmission of images from Washington, DC, to the auditorium of the Bell Telephone Laboratories in New York, a distance of over 250 miles. During the radio demonstration, images were sent from the Bell Laboratories experimental station 3XN at Whippany, New Jersey, to New York City, a distance of 22 miles. Reception was by means of two forms of receiver. One receiver produced a small image of approximately 2.0 inches × 2.5 inches, which was suitable for viewing by one person. The other receiver gave a large image of nearly 24 inches × 30 inches for viewing by an audience of considerable size.

Ives and his colleagues used a Nipkow disc with fifty apertures for scanning purposes. They arrived at this figure by taking as a criterion of acceptable image quality the standard of reproduction of the half-tone engraving process in which it was known that the human face can be satisfactorily reproduced by a fifty-line screen. Thus, assuming equal definition in both scanning directions, 2,500 elements or 40,000 elements per picture had to be transmitted for a rate of picture transmission of sixteen pictures per second. The frequency range needed to transmit this number of elements per second was calculated to be 20 kHz.

A spotlight scanning method[25] was adopted to illuminate the subject, the beam of light being obtained from a 40 A Sperry arc. Three photoelectric cells of the potassium hydride, gas-filled type were specially constructed and utilized to receive the reflected light from the subject. At the time they were probably the largest cells that had ever yet been made and presented an aperture of 120 inches.[26]

For reception a disc similar to that at the sending end was used together with a neon glow lamp. The disc had a diameter of 36 inches and synthesized the 2.0 × 2.5 inch image. Another form of receiving apparatus comprised a single, long, neon-filled tube bent back and forth to give a series of fifty parallel sections of tubing. The tube had one interior electrode and 2,500 exterior electrodes cemented along its rear wall. A high-frequency voltage applied to the interior electrode and one of the exterior electrodes caused the tube to glow in the region of that particular electrode. The high-frequency modulated voltage was switched to the electrodes in sequence from 2,500 bars on a distributor with a brush rotating synchronously with the disc at the transmitting end. Consequently, a spot of light moved rapidly and repeatedly across the grid in a series of parallel lines, one after the other, and in synchronism with the scanning beam. With a constant exciting voltage the grid appeared uniformly illuminated but when the high-frequency voltage was modulated by the vision signals an image of the distant subject was created. To transmit the vision, sound and synchronizing signals three carrier waves were employed: 1575 kHz for the image signals, 1450 kHz for the sound signals and 185 kHz for the synchronization controls (Figure 2).[27]

According to Ives, the success of the system was due to the 'chief novel features' listed below:[28]

1. The choice of image size and structure such that the resultant signals fell within the transmission frequency range of the available transmission channel.
2. The scanning by means of a projected moving beam of light.
3. The transmission of only the a.c. components of the image.
4. The use of self-luminous surfaces of high intrinsic brilliancy for the reconstruction of the image.
5. The utilization of high-frequency synchronization.

The first demonstration consisted of the transmission of an image of, and an address by, Herbert Hoover, Secretary of Commerce, from Washington to New York over telephone lines. The second demonstration by radio comprised three events: first, an address by E.L. Nelson, a Bell Laboratories engineer; second, a 'vaudeville act' featuring 'a stage Irishman, with side whiskers and a broken pipe, [who] did a monologue in brogue', and then, after a quick change, returned with a blackened face and made a few quips in dialect; and, finally, a short humorous dialect talk.[29]

The received images were subject to some fading and ghosting and occasionally appeared in the negative but in general they impressed the audience. 'It was as if a photograph had suddenly come to life and began to talk, smile, nod its head and look this way and that,' said one observer.

Colonel Angwin, the deputy Chief Engineer of the UK's General Post Office witnessed a demonstration of the Bell Laboratories television system some time after the public demonstration.[30] In his report he mentioned: 'This system reproduces a clear and undistorted picture and the results obtained are undoubtedly far in advance of those claimed for and by the Baird system. The American system is a very costly and elaborate piece of mechanism and requires a special circuit for line transmission and exceptionally stable conditions for wireless transmission.' Angwin's visit occurred about one year later than the April 1927 trial but during that time nothing had been done by the company to exploit the system commercially and Angwin was given to understand that it had no intention of proceeding further with it at that time. Dr Dauvillier, the eminent French physicist, observed the wireless transmission and, in a historical review article on television published in the *Revue Générale de l'Électricité*, wrote:

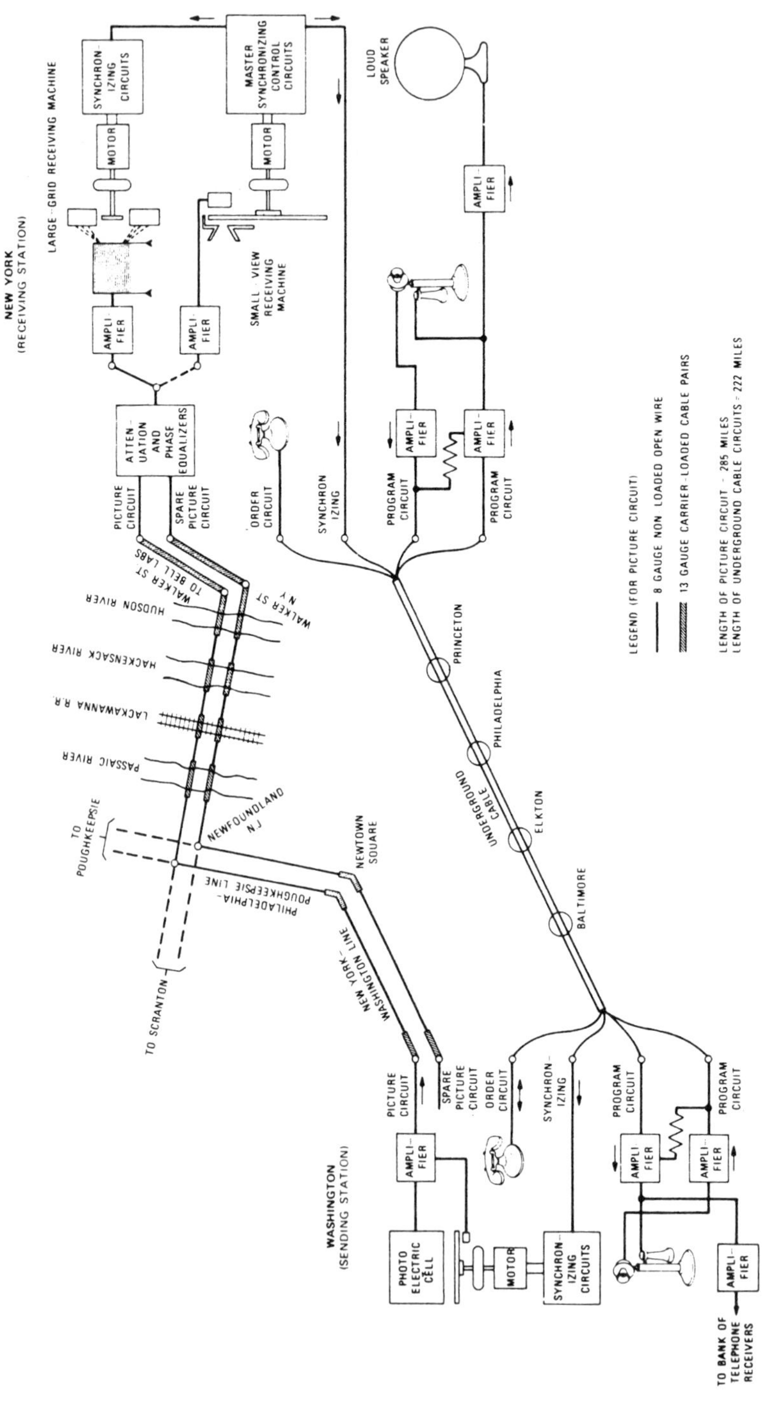

Figure 2 Schematic diagram of circuits for 1927 television demonstration.

'Finally, the Bell Telephone Company recently succeeded in transmitting to a considerable distance the human face, using (without acknowledgement) the Baird system.'[31]

The April 1927 demonstrations were the finest that had been given anywhere even though no especially novel features had been incorporated into the various systems. They established standards from which further progress could be measured. Moreover, the publication in October 1927, in the *Bell System Technical Journal* (Volume 6, pp. 551–653), of five detailed papers on the factors which led to Ives's group success enabled other workers to ponder on whether their own ideas and practices were likely to lead to similar favourable outcomes. Certainly, the Bell Laboratories equipment could be further developed. The large screen grid display was 'very much inferior' to that of the gaseous discharge lamp; the person being televised had to sit in a semi-darkened room; there was a need to dispense with the separate synchronizing channel; and there was a requirement for more detailed images. In addition to these considerations the policy of AT&T towards television advancement in the Bell System had to be defined.

Ives felt there were three principal projects to be tackled.[32] First, the introduction of a two-way appointment service between New York and Washington, 'as a means of keeping the Bell System on the map in connection with the onward course of television, while at the same time securing information as to its possible uses and problems'. Second, transatlantic television would be the 'supreme achievement' in television and would have 'an appeal to the imagination of all ranks of humanity which would be unsurpassable'. Third, the development of the public address television apparatus, and its possible use for televising a presidential inauguration, would find its justification if it were the 'policy of the Telephone Company to either provide a service of this sort or through its subsidiaries to manufacture apparatus from the sale of which, or from the use of which, income could be expected'.

Of these proposals, approval was given for the first and third. The second was thought to be a publicity affair.[33] F.B. Jewett, President of Bell Telephone Laboratories, believed that Ives's group should proceed as vigorously as possible with the preliminary work of the third project, without there being any definite commitment. In addition the Laboratories should carry on, as adequately as possible whatever fundamental work would be necessary to safeguard the company's position and advance the art along lines that were likely to be of interest to the company. This mandate gave Ives ample scope to investigate a quite wide range of television problems. He seized the opportunities made accessible to him and during the next three years daylight television, large-screen television, television recording, colour television and two-way television were all subject to his group's scrutiny and engineering prowess.

The televising of objects illuminated by natural daylight by the method of direct scanning was demonstrated on 10 May 1928. Studies of the optical conditions peculiar to television had brought out the simple fact that the light-gathering power of a lens and television scanning disc could be increased by enlarging the physical dimensions of the whole scanning system.[34] Other work in the Laboratories had led to the development of photoelectric cells of greatly increased sensitivity. With these cells, of the thalofide type, a fifty-hole, 36-inch diameter scanning disc and a lens system of aperture *f*/2.5, it was practicable to transmit images of a full-length human figure when the apparatus was taken out of doors and set up on the roof of the Bell Laboratories building in New York. A press show was given on 12 July 1928.[35] This contained scenes of a sparring match, a golf exhibit and other movements.

The evolution of equipment for public address television proceeded along several paths. First, in the grid display receiver of 1927 the form of brush used and the method of making contact with the individual electrodes on the grid were much improved: Ives could 'guarantee', in February 1928, that a face could be reproduced so as to be 'very satisfactorily recognizable'.[36] Second, in 1927 Hartley and Ives patented methods for projecting televised images by photographing the received image with a cine camera and developing the film images with the minimum of delay.[37] They also patented the generation of television signals from rapidly processed cine film. The advantages of using motion picture film at the sending end of a television system were known to Ives in 1925. In his January 1925 memorandum to Arnold,[38] Ives referred to measurements which he had made of the brightness of the image of a sunlit landscape as projected on to a photo-electric cell by means of a wide-aperture lens. The measurements revealed that the magnitude of the light flux which could be concentrated on the cell was about 0.2% of that employed in picture transmission. Hence the degree of sensitivity of the photocell and the degree of amplification necessary in the proposed television system would be far greater than those that had been acceptable in connection with picture transmission.

Ives advocated using transmitted light (from a film), rather than reflected light (from a scene), to ease the solution of the problem.[39] He noted: 'It may be pointed out that the use of a moving picture film as the original moving object is equivalent to a very great amplification of the original illumination brought about by the photochemical amplification process involved in the production and subsequent development of the photographic latent image.' This intermediate-film system of television was employed in the 1930s by Fernseh AG and Telefunken in Germany, and by Baird Television Ltd at the London Television Station, Alexandra Palace.[40]

Third, Gray devised a method of projecting television images based on the optical projection of a small, illuminated and moving section of a slot cut in a rotating disc placed in front of a capillary light source.[41] This lamp could operate either as a glow discharge through mercury vapour or as a mercury arc. In February 1929 Gray could exhibit fifty-line, 21 × 32 inch projected television images for viewing in a darkened room and 11 × 14 inch screen images for viewing in a room only partially darkened. A modification of this apparatus allowed fifty-line television images at eighteen images per second to be recorded on 35 mm motion picture film.[42]

When Ives was admitted in 1905 to Johns Hopkins University, as a PhD student, it was to study colour photography under the supervision of R.W. Wood, then the leading authority on optics in the United States. Ives's choice of subject had possibly been influenced by his father's contributions to the art of photography and the science of optics. Dr F.E. Ives had invented (among other things) a trichromatic camera, various processes of colour photography and a 'device for optically reproducing objects in both full modelling and natural colours.' H.E. Ives's first two papers (both published in *The Physical Review*) relate to improvements in methods of colour photography; the earlier paper (1906)[43] pertains to Wood's device and the later paper (1907)[44] to Lippmann's scheme. The papers were written before Ives had completed his doctoral thesis (1908), which had the title 'An experimental study of the Lippman colour photograph'.

With such a background it was perhaps inevitable that Ives and his co-workers would want to attempt to create coloured televised images. On 27 June 1929 colour television was shown by Bell Telephone Laboratories to an invited gathering of scientists and journalists.[45] Ives utilized three signal channels so that the three colour signals could be sent simultaneously from the transmitter to the receiver.[46] An advantage of this arrangement was that the same scanning discs and motors, synchronizing equipment and light sources, and the same type of circuit and method of amplification were used as in the monochrome scheme. The only new features were the form and disposition of the specially devised photocells at the sending end and the type and grouping of the neon and argon lamps at the receiving end (Figure 3).

A neon glow lamp gave the desired red light but for the sources of green and blue light 'nothing nearly so efficient as the neon lamp was available'. Two argon lamps, one with a green filter and one with a blue filter, were finally adopted for the demonstration; however, various expedients were needed to increase their effective luminous intensity. Special lamps with long, narrow and hollow cathodes cooled by water were utilized and these were observed end-on so that the thin glowing layer of gas was greatly foreshortened and the apparent brightness thereby increased. To render the correct tone of coloured objects it was essential to obtain photoelectric cells which would be sensitive throughout the visible spectrum. A.R. Olpin and G.R. Stilwell constructed a new kind of cell which used sodium in place of potassium. Its active surface was sensitized by a complicated process involving sulphur vapour and oxygen, instead of by a glow discharge of hydrogen as with the former class of cell.

An account of the demonstration, in which the transmission was over lines, was published in *Telephony* on 6 July 1929.[47] The display

> opened with the American flag fluttering on a screen about the size of a postage stamp. The observer saw it through a peephole in a darkened room. The colours reproduced perfectly. Then the Union Jack was flashed on the screen and was easily recognized by its coloured bars.
>
> The man at the transmitter picked up a piece of watermelon, and there could be no mistake in identifying what he was eating. The red of the melon, the black seeds and the green rind were true to nature, as were the red of his lips, the natural colour of his skin and his black hair.

Figure 3 Bell Laboratories apparatus for colour television. With the exception of the photoelectric cabinet on the left, the apparatus is identical with that used for the original demonstration of monochromatic television.

As previously noted, of the two projects approved by Jewett one pertained to two-way television. This was established between the main offices of the AT&T Company at 195 Broadway, New York, and the Bell Telephone Laboratories at 463 West Street, New York, and was demonstrated on 9 April 1930.[48] It consisted essentially of two complete television transmitting and receiving sets of the kind employed in the 1927 one-way television scheme. Spotlight scanning, Nipkow discs and neon lamps were still incorporated but with several improvements. Two discs, each containing seventy-two holes to give double the image detail as compared with the fifty-hole discs of the 1927 apparatus, were utilized at each end of the line links, one for image analysis and the other for image synthesis. In addition to the photoelectric cell and neon lamp each 'ikonophone' booth had a concealed microphone and loudspeaker. Special precautions had to be taken with the telephone circuit to prevent 'singing' due to the closeness of the two electro-acoustic transducers. The increased bandwidth of the system led to problems of amplitude and phase equalization which were more difficult than those encountered in the earlier tests.[49]

On the optical side[50] the principal problem was that of regulating the intensity of the scanning light and of the received image so that 'the eye [was] not annoyed by the scanning beam or the neon lamp image rendered difficult of observation'. Ives and his colleagues, Gray and Baldwin, solved this difficulty by using a scanning light colour to which the eye is relatively insensitive, but to which the photoelectric cells could be made highly sensitive.

When the two-way system was withdrawn from service it had been seen by more than 17,000 people. A novel application was observed when two deaf persons carried on a telephone conversation by reading each other's lips.[51] The cost of providing the service, by the New York Telephone Company, was estimated to be $15,350 per year, excluding the cost of the technical operation and maintenance, which was borne by Bell Telephone Laboratories.

Much publicity was given to the demonstrations: for the first six months of 1931 more than 700 column-inches of newsprint were devoted to reports in the city papers.[52] Rather oddly, perhaps, the desirability of this type of publicity was questioned by the Administration Department of the AT&T Company. 'Our exhibit emphasizes to me that commercial use of television is a long way off,' wrote a member of the department in March 1931.[53]

Following the completion of the two-way link Ives undertook an important appraisal of the progress which had been made by his group and endeavoured to define the course of action that had to be implemented for the future advancement of television. His prognosis was gloomy in outlook.[54] For Ives the statement of the problem that had to be solved was simple:

> An electrically transmitted photograph 5 in by 7 in in size, having 100 scanning strips per inch, has a field of view and a degree of definition of detail, which, experience shows, are adequate (although with little margin) for the majority of news events pictures. It is undoubtedly a picture of this sort that the television enthusiast has in the back of his mind when he predicts carrying the stage and the motion picture screen into the house over electrical communications channels.

The difficulty of achieving this desirable result was readily apparent. In the photograph the number of picture elements is 350,000, and at a repetition speed of twenty per second (twenty-four per second had now become standard with sound films) this meant the transmission of 7,000,000 picture elements per second and a bandwidth of 3.5 MHz for the system on a single sideband basis. Ives compared the criteria for high-definition television and the results which had been obtained in America, and observed: 'All parts of the television system are already having serious difficulty in handling the 4000-element image.' (This was the number of image elements used in the seventy-two-line picture of the two-way television link.)

The obstacles that had to be overcome before a high-definition system could be implemented were found in the use of the scanning discs at the transmitter and receiver, the photoelectric cells, the amplifying systems, the transmission channels and the receiving lamps. Ives noted that the disc, while quite the simplest means for scanning images of few elements, was entirely impractical when really large numbers of image elements were in question, and wrote: 'As yet however, no practical substitute for the disc of essentially different character has appeared.'

Turning next to photocells, there were, in 1930, two types of cell which could be utilized for television: the gas-filled cell, which had a good sensitivity but poor frequency response; and the vacuum cell, which was much less sensitive than the gas-filled cell, although it was free from its failing. The self-capacity of the cells and the associated wiring and amplifier caused the high frequencies to be attenuated relative to the lower frequencies and consequently equalizing circuits with their attendant problems of phase adjustment, together with more amplification, were needed. But Ives observed that amplifiers capable of handling frequency bands extending from low frequencies up to 100,000 Hz or more gave serious problems.

The communication channels, either radio or wire, also posed grave difficulties for high-definition television and its related bandwidth specification.

> In radio, fading, different at different frequencies, and various forms of interference stand in the way of securing a wide frequency channel of uniform efficiency. In wire, progressive attenuation at higher frequencies, shift of phase, and cross-induction between circuits offer serious obstacles. Transformers and intermediate amplifiers or repeaters capable of handling the wide frequency bands here in question also present serious problems.
>
> Finally at the receiving end of the system the neon glow could not follow satisfactorily television signals well below 40 000 Hz, and, in the case of the 4000-element image the neon had to be assisted by a frequently renewed admixture of hydrogen, which again could not be expected to increase the frequency range indefinitely. With the receiver disc, as at the sending end, increasing the number of image elements rapidly reduced the amount of light in the image and, with a plate glow lamp of given brightness, the apparent brightness of the image is inversely as the number of image elements.

These considerations led Ives to one clear conclusion: 'The existing situation is that if a many-element television image is called for today, it is not available, and one of the chief obstacles is the difficulty of generating, transmitting, and receiving signals over wide frequency bands.' A partial solution was to employ multiple scanning and multiple-channel transmission. The beginnings of such an approach had been given by P. Mertz in his 1921 memorandum.[55] Ives's multi-channel experimental set-up is illustrated in Figure 4. This used scanning discs with prisms over their apertures, so that, at the sending end, the beams of light from the successive holes were diverted to different photoelectric cells. At the receiving end, the prisms enabled beams of light from the three lamps to be deflected in a common direction.

Ives found that his three-channel apparatus yielded results strictly in agreement with the theory underlying its conception and observed that the 13,000-element image was a marked advance over the single-channel 4,000-element image.

> Even so, the experience of running a collection of motion picture films of all types is disappointing, in that the number of subjects rendered adequately by even this number of image elements is small. 'Close-ups'

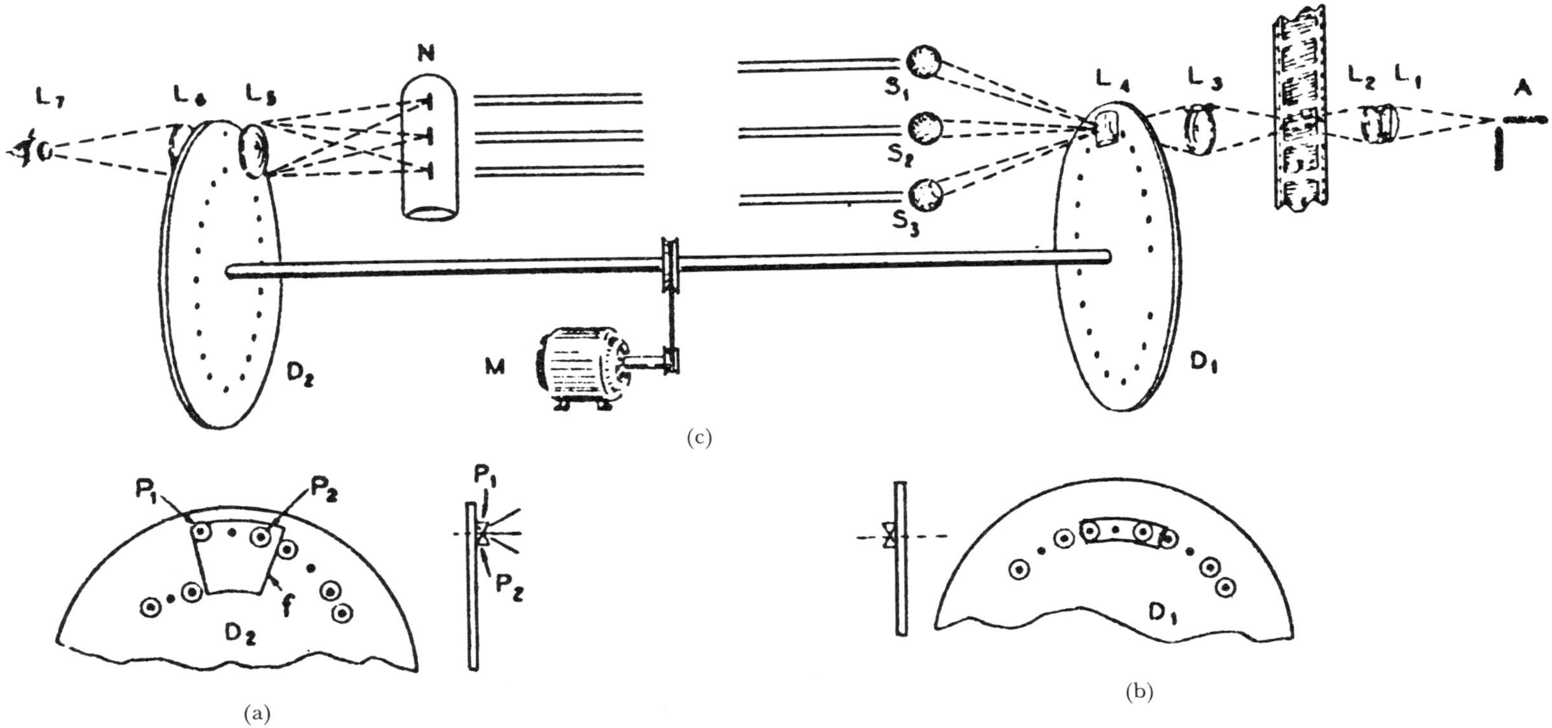

Figure 4 Schematic of three-channel television apparatus. (a) Receiving end disc with spiral of holes provided with prisms. (b) Sending end disc with circle of holes provided with prisms. (c) General arrangement of apparatus.

> and scenes showing a great deal of action, are reproduced with considerable satisfaction, but scenes containing a number of full length figures, where the nature of the story is such that the facial expression should be watched are very far from satisfactory. On the whole the general opinion . . . is that an enormously greater number of elements is required for a television image for general news or entertainment purposes.

Interestingly, multi-channel television schemes were also demonstrated in January 1931 by the Gramophone Company and by Baird Television Ltd.[56]

The points made by Ives in his appraisal of the generation and display of video signals had been appreciated by a few workers for several years. Campbell Swinton, in his Presidential Address of 1911 to the Roentgen Society,[57] had commented upon the impractical nature of the pursuit of perfection by mechanical means and had outlined an all-electronic system of television. Zworykin had patented a version of such a system in 1923 and when, in 1925, Farnsworth started his work on television it was on the basis of an all-electronic scheme.[58] Again, in Great Britain, a similar approach was to be adopted in 1931 by Electric and Musical Industries Ltd.[59] Subsequently, Zworykin, Farnsworth and EMI satisfactorily demonstrated all-electronic television before the end of 1935.

The year 1931 marked a turning point in the progression of television at the Bell Telephone Laboratories. Following a period of six years (1925–30) as a centre of excellence the television laboratories declined in importance. This decline stemmed from certain constraints that had been imposed on the business of the AT&T company and also from the certain lack of direction of the R&D television effort.

In May 1931 Ives wrote a memorandum on a 'Future program for television research and development'.[60] At that time the Laboratories' television projects were not aimed at any specific developments, such as radio or wire broadcasting into theatres or homes, the improvement of large images, or small receivers for home use. Their work was mainly exploratory in the realm of fundamentals. Ives's conviction was that the only real and lasting field for television, if it could be developed, was that of home entertainment. Consequently, Ives felt that the only worthwhile problems in television research were the improvement of broadcasting methods and the advancement of terminal apparatus, especially receiving apparatus, to produce the high-grade image which alone would be satisfactory. He realized the magnitude of the task required and wrote: 'The technical difficulties are enormous and the possibility of overcoming them is veiled in obscurity.' Nevertheless, these problems could be faced with interest and enthusiasm by his staff but 'a damper [was] put on thought and planning along these lines by the knowledge that television for home entertainment [lay] outside the chosen sphere of activities of the Bell System. It was debarred therefrom by its present contract relations.'

On 1 July 1926, AT&T and RCA had signed an agreement according to which, for $1,000,000, the AT&T company had transferred its radio facilities (such as station WEAF) to RCA and had withdrawn from broadcasting. It had also given up its rights to manufacture receiving sets in favour of RCA. As a *quid pro quo*, RCA had agreed to use the Bell System's wire network exclusively and not to compete with the telephone company for telephone business.[61] Ives was thus faced with a dilemma. Either a drastic decision to drop the whole subject of television as one foreign to the Bell System could be taken, or a programme of research, on both sending and receiving equipment, could be initiated as though the whole field was open to the Laboratories, in 'the hope or with the definite intention of so changing [the] policy and contract relations that [the company would] ultimately go into [this] field'.

The position relating to television enterprises and publicity was a sensitive one *vis-à-vis* the AT&T–RCA contract. Caution in publicity matters was exercised and in 1931 no announcement of television progress was given in the Annual Report of the AT&T Company. There was a reference to cables which would permit an extremely wide band of frequencies to be transmitted, but without any statement of its application in a television system. No mention of either wideband cables or television was given in the 1932 and 1933 Reports: they noted the depressed business conditions and the Depression respectively. Thereafter nothing was said, in the prewar Annual Reports, of the company's television ventures *per se*.[62] Again, after 1932 the editor of *The Telephone Almanac*, R.T. Barrett, was required to delete from future issues of the journal all references to the Bell System's television activities.[63] Publicity for these, in *The Telephone Almanac*, was not renewed until 1946. The lack of public disclosure did not denote a cessation of work on television. On the contrary, from 1931 to 1935 (inclusive) the expenditure of the company on television advancement amounted to $592,400.[64]

During its programme of work on television Ives's group had undertaken some investigations (in 1926) on the appropriateness of utilizing a cathode ray tube in a television receiver.[65] Images of simple objects, such as a letter A, a bent wire, etc., had been received on a modified cathode ray tube. The received picture signals had been impressed on an extra grid in the tube and controlled the intensity of the electron beam incident on the fluorescent screen. In one type of tube, the grid was close to the hot filament and the picture signals acted on the beam before it reached the accelerating field. In a second type of tube, the cathode beam passed through two parallel wire gauzes just before striking the screen and the image signals were applied across these two gauzes.

Both types of tube reproduced images of the simple objects mentioned above but they did not reproduce images of more complex objects, nor did they show half-tones in a satisfactory way. The difficulty was that a change in intensity of the cathode beam from its normal value altered both the focus of the beam and its point of incidence on the screen. These two factors resulted in a serious distortion of the synthesized image. The problems associated with gas-filled cathode ray tubes were to become known to other experimenters, including Farnsworth and Zworykin, and so endeavours were made to evolve vacuum cathode ray tubes. Some later workers, for example Bedford and Puckle, and von Ardenne, sought to ameliorate the effects of gas-filled tubes by adopting velocity modulation.[66]

On 18 November 1929 Zworykin gave a lecture on his new 'kinescope' or picture tube, to a meeting of the Institute of Radio Engineers. His efforts to develop a vacuum picture receiving tube had been successful and represented an important step forward in the slow march towards an all-electronic television system. Ives was not impressed though. A demonstration of the kinescope had not been given at the IRE meeting and the account of Dr Zworykin's work was 'chiefly talk'. 'This method of reception,' wrote Ives, 'is old in the art and of very little promise. The images are quite small and faint and all the talk about this development promising the display of television to large audiences is quite wild.'[67]

With the leader of the television group holding such an opinion it was perhaps inevitable that R&D effort on the evolution of an all-electronic television scheme would not be a major part of the group's activities. Instead, in his May 1931 memorandum on 'The future programme for television research and development',[68] Ives confirmed his support for mechanical scanning by suggesting three special projects all based on this mode of reconstituting an image. These were: (1) the demonstration of reception from an aeroplane; (2) the demonstration of direct scanning of some major outdoor event; (3) the demonstration of reception on film and thence projection in the theatre. All these recommendations if approved would have been based on existing principles and technology. Ives did not propose the investigation of electronic cameras or electronic receiving tubes, and yet when his group eventually embarked on such a programme an important advance was made.

Essentially, by 1931 the analysis and synthesis of objects and images by mechanical scanning had shown itself to be unsuitable for the reproduction of high-definition images. This inappropriateness was to be highlighted in 1936, when the BBC's London television station was established. Initially two alternative services were provided by the BBC, utilizing equipment manufactured by Marconi–EMI Television Ltd and by Baird Television Ltd respectively. The Marconi–EMI apparatus was wholly electronic, whereas that of the Baird company employed mechanical scanners for its spotlight camera, intermediate film studio camera and telecine apparatus. Within only two months of the opening of the London station the BBC's studio staff and engineers had decided that the Baird systems, with the exception of the telecine scanner, could not compete on equal terms with the emitron cameras of Marconi–EMI. From February 1937 only that company's equipment formed part of the UK's television service.[69]

The disappointment suffered by Baird resulted from his ignoring for too long the inevitable move towards high-definition television and the use of cathode ray tubes in receivers and cameras. In October 1931 he was quoted as saying that he saw 'no hope for television by means of cathode ray bulbs', that 'the neon tube [would] remain as the lamp of the home receiver', and that he was sceptical about the success of the use of short waves in television 'because they covered a very limited area'.[70] The early successes of Ives's BTL group mirrored those of Baird and his collaborators, although they were achieved quite independently. Both BTL and Baird embarked on their pursuits by adopting the Nipkow disc (and its variants) as the principal scanning element in their schemes. They both adapted their basic system to demonstrate spotlight scanning, large-screen television, daylight television, colour television, two-way television and multi-channel television, *inter alia*. Moreover, the procedures which were perforce necessary to accomplish some successes were similar (see Table 3).

By 1931 it was evident that new approaches to the problem of seeing by electricity were needed if television broadcasting was to become widespread in popular appeal. This appeal would be encouraged when the Derby, tennis at Wimbledon, sporting and athletic events, news reports and so on could be televised to give adequate image detail.

Zworykin, Shoenberg and Farnsworth, among others, were cognizant of this point and worked strenuously to perfect their companies' electronic camera tubes: the iconoscope, emitron and image dissector respectively. Zworykin described, but did not demonstrate, his ideas at a meeting of the Institute of Radio Engineers held on 26 June 1933 in Chicago. The meeting was attended by Dr F. Gray, of Ives's group. He wrote a short memorandum (dated 6 July 1933)[71] on what he had learned from the lecture and on the same day Ives sent a memorandum to Dr H.D. Arnold,[72] the Director of Research at BTL. 'The device,' wrote Ives, 'involves some principles which are new in the television art and represent a considerable advance in the direction of attaining a television transmitting device suitable for general use in such places as a moving picture camera might be used.' He felt Zworykin had taken an important step in the right direction and opined: 'it is not at all improbable that there will emerge from his development a television transmitter which will make high grade transmission of television material suitable for entertainment a practicable possibility'.

Several days later (20 July 1933) Ives, in a short memorandum,[73] noted: 'The consensus of opinion at our conference yesterday [on the subject of cathode ray transmitting apparatus] appears to be the feeling that we would do well to start some work along these lines.' On 31 July Gray outlined a possible programme of work and soon a beginning was made on an enquiry into the design of an electronic camera tube.[74] Unfortunately for the prestige of Bell Telephone Laboratories this effort began three years too late. On 20 May 1930 Gray had written a comprehensive memorandum (twenty-one pages long and with eleven figures) on a 'proposed television transmitter'[75] based on charge storage, in which he had put forward eleven methods by which a vision signal might be generated by various photoelectric means. One of his objects in listing the proposals was 'to aid in deciding the direction of certain future researches on television'. Had Gray's inventiveness and resourcefulness been acted upon it is possible the Laboratories, with their excellent experimental facilities and staff, would have demonstrated an electronic camera contemporaneously with Zworykin's demonstration.

Ives's team seems, however, to have been reinvigorated by its new task. Within a year: Gray had proposed a cathode ray transmitter that utilized

Table 3 Some contributions (by Bell Laboratories and Baird) to the development of television in the 1920s

Use of/demonstration of	*Baird*	*Bell Laboratories*
1. Nipkow discs	From 1923	From 1925
2. Glow discharge lamps	Various experiments 1924 on	Various experiments 1925 on
3. Means to reduce time lag of photocells	From *c.* 1925. Used derivative of photocell current. Patent 270 222, 21 October 1925	From *c.* 1925–6. Used C–R coupling circuit to enhance high-frequency gain. Internal memorandum 27 February 1926
4. Coloured filters on lamps to reduce discomfiture of persons being televised	Various experiments in 1925. Demonstration of infra-red radiation, 23 November 1926	Various experiments in 1925. Mentioned in internal memorandum, 26 August 1925
5. Large Nipkow discs	Utilized discs up to 8 feet (2.44 m) in diameter at some time during the period 1923–5	Advantage of using discs up to 10 feet (3.05 m) in diameter mentioned in an internal memorandum, 27 July 1925
6. Spotlight scanning	Employed from 1926 to 1936. Patent 269,658, 20 January 1926	Employed from *c.* 1925–6. US patent applied for on 6 April 1927. UK patent 288,238, 18 January 1928
7. Two-way television	Patent 309,965, 19 October 1927	UK patent 297,152, 17 June 1927. Demonstrated from 9 April 1930 to 31 December 1932
8. Transatlantic television	Demonstrated 9 February 1928	Suggested as a publicity event in an internal memorandum 4 May 1927
9. Intercalated images to improve resolution	Patent 253,957, 1 January 1926. Various experiments *c.* 1924	Mentioned in internal memorandum 9 September 1927
10. Colour television	Demonstrated 3 July 1928	Demonstrated July 1928
11. Large-screen television	Demonstrated 28 July 1930	Demonstrated 7 April 1927
12. Daylight television	Demonstrated June 1928	Need to work on natural-light scanning mentioned in an internal memorandum 4 May 1927. Demonstrated 16 July 1928
13. Commutated lamp bank/display	Patent 222,604, 26 July 1923. Demonstrated 28 July 1930	Demonstrated 7 April 1927
14. Use of arcs	Demonstrated January 1931	Mentioned in internal memorandum 11 February 1929. Experiments 1929
15. Zone television	Demonstrated 2 January 1931. Patent 360,942, 6 August 1930	Described in a paper by H.E. Ives, 'A multichannel television apparatus', *J. Opt. Soc. Am.*, 1931, 21. Demonstrated 1930–1

the stored charge flowing through a photoconductive film during a complete image cycle,[76] and had put forward various ideas for projecting images from the screen of a cathode ray tube, whose optical reflecting power and transparency could be affected by the electron beam[77] (cf. the Eidophor); Teal had reported on experiments undertaken with new photoelectric emitters (Cs-Cs_2S-Ag and K-K_2O-Ag surfaces) suitable for use in an iconoscope;[78] a conference had been held at which seven novel methods for displaying large images in theatres had been discussed;[79] Ives had suggested a very simple form of iconoscope suited for generating picture signals from continuously moving motion picture film;[80] and Hefele had demonstrated a basic form of electronic camera.[81]

By the end of July 1934 photosensitive targets, of the charge storage type, had been prepared by two methods and tested. In one form, one side of a very thin sheet of mica had been coated with metal to form one plate of a capacitor, while the other plate (the reverse side of the mica sheet) had consisted of tiny globules of pure sodium deposited by carefully controlled distillation. The other method consisted of ruling a sputtered platinum film, on the face of a mica sheet, into small elements 0.004 inches square. A very thin coating of pure sodium was then allowed to be deposited on to these platinum squares to enable each one to be photosensitive.

A specially constructed cathode ray tube was used with these targets. Hefele reported: 'Shadowgraphs of various objects focussed onto the sensitive plate of the transmitting tube were clearly recognized. . . . These encouraging results mark a forward step in our study of non-mechanical television transmission. Attention is being paid to the development of higher sensitivity photo-electric mosaics and to the enlargement of picture detail by the transmission of more picture elements which involves research in cathode ray structures as well as amplification and transmission of a very wide frequency range.' Twelve months later, on 26 July 1935, Hefele demonstrated to several senior members of the Bell Laboratories a seventy-two-line photoconducting television camera tube. It contained a film of red mercuric iodide, cooled by solid carbon dioxide to render the film stable and so permit easier working conditions over long periods of time. The electron gun of the tube was designed by Gray and the received images were displayed on a cathode ray tube of C.J. Davisson's design. In view of the 'excellent images' obtained, it was decided to attempt 'immediately' to construct a 240-line camera tube.[82]

By 6 August 1935 Davisson had submitted a design for a 240-line electron gun (of 5 μA beam current) to give a spot size of 0.008 inches diameter so that an existing 2 × 2 inch photoconducting surface could be employed. Messrs Jensen and Strieby had promised to provide two 240-line sweep circuits by 1 September.

Hefele's demonstration appears to have been the first ever given of a photoconductive camera tube. The demonstration showed that Bell Telephone Laboratories had the expertise, the experimental facilities and the essential ideas to engineer an all-electronic television system. Nevertheless, their accomplishments were in some respects approximately two years behind those which had been obtained by RCA and EMI. The period of comparative inactivity of Ives's group (January 1931 to July 1933) had put it at a disadvantage relative to Zworykin's team at RCA and Shoenberg's group at EMI. At both RCA and EMI work on an all-electronic television system was being pursued with some zeal, notwithstanding the depressed business conditions.

Following the establishment of EMI Ltd in the United Kingdom in 1931, the Director of Research, Isaac Shoenberg, rapidly created a powerful research group to engage in television development work.[83] By June 1934 the research department comprised thirty-three university graduates (many of whom held doctorate degrees), twenty-eight laboratory assistants (of sub-BSc standard), seven draughtsmen, thirty-three mechanics and glassblowers, and seven women, a total of 108 personnel. At this time the company was spending about £100,000 per year on television research and development. This effort led Shoenberg and his colleagues, on 15 February 1935, to advance their company's proposals for high-definition television at a meeting of the Technical Sub-committee of the Television Advisory Committee. These were based on 405 lines per picture, twenty-five pictures per second and fifty frames per second—a standard which had never been put forward or demonstrated previously anywhere in the world. During the second week of April in 1935 EMI was able to demonstrate the work to Gerald Cock, the BBC's Director of Television: he was very impressed. Public showings of the system were given at the Radio Exhibition at Olympia, London, from 26 August 1935 and the world's first regular, public, all-electronic, high-definition system of television commenced broadcasting on 2 November 1936.

The convictions of Zworykin, Farnsworth, Sarnoff, Shoenberg and others concerning the inevitability of television broadcasting did not seem to be shared by Dr F.B. Jewett, the President of Bell Telephone Laboratories. During his testimony before the Federal Communications Commission (the Hearing of 10 December 1936) he expressed himself as being pessimistic on television matters and stated: 'I am very sceptical personally about television. . . . Now, if it does come—and I am just as clear as a crystal in my own feelings on this thing—if it comes, if these people who are interested in it are right in their belief, and I am wrong . . .'[84] None the less, work proceeded at BTL and on 9 November 1937 the Laboratories demonstrated the transmission of video signals along a coaxial cable which had recently been installed between New York and Philadelphia, a distance of approximately 200 miles.[85]

The patent for the cable had been filed on 23 May 1929 by Espenschied and Affel, and mentioned that one objective was the use of the wideband long-distance transmitting medium for television transmission. Prior to the New York–Philadelphia installation, the cable had entered service on 10 June 1936, providing the means for propagating signals from NBC's studio in New York to the transmitter in the Empire State Building, a distance of 1.5 miles. The intended use of the new cable system was, partly, the simultaneous transmission of 240 separate signals, each requiring a 4 kHz bandwidth, of a carrier channel telephony system. Testing of the 960 kHz bandwidth of the cable using one signal demanded the use of a video signal generator. For this purpose a mechanical scanner and motion picture film projector formed the basis of the test equipment.

The six-foot-diameter steel disc, driven by a 10-horsepower motor, incorporated 240 large-aperture lenses, each located at the same distance from the centre of the disc, and rotated at 1,440 r.p.m. (i.e. 24 revolutions per second). The lenses focused a light beam to give a square scanning spot on the film of size 0.003 × 0.003 inches. After passing through the film the light was incident on a photomultiplier and the output from it was applied to a modulator which shifted the band of picture frequencies from 0–800 kHz to 100–900 kHz, the region between 0 and 100 kHz being utilized for the sound and synchronizing channels.

The receiver employed a specially designed cathode ray tube, known as a Davisson tube, and gave an image of 8 × 10 inches. Although there was some flicker and the brightness of the image was low, it was reported that the picture was of high quality. 'The outstanding characteristics of the image were the crispness of the detail along each line, the sharp demarcation and detail in the shadows, and freedom from phase "hang-over" effects.'[86] These desirable features stemmed from the decision not to use an iconoscope and interlaced scanning. Both RCA and EMI employed these in their television systems but the use of the iconoscope or emitron led to 'shading' problems at the receiver, and interlaced scanning could lead to a loss of image detail if there were imperfections in the timing of the line sweeps. The BBC, too, found that the telecine mechanical scanner of Baird Television Ltd gave higher-quality images than the emitron scanner of EMI.[87]

Three years after the test, on 21 May 1940, 441-line video signals with a bandwidth of 2.7 MHz were sent along the coaxial cable from New York to Philadelphia and back. Demonstrations of such transmissions were subsequently given before the National Television System Committee (NTSC) on 8 November 1940, and before the Institute of Radio Engineers in January 1941. Previously, on 20 May 1939, standard telephone cable pairs had been used successfully as local transmission lines for the broadcasting of a six-day cycle race at Madison Square Garden, New York.[88] These were the last prewar demonstrations of television given by Bell Telephone Laboratories. During 1940 television activity ceased, increased effort thereby being made available for defence work.

Of all the contributions made by the staff of the Laboratories in the field of television, perhaps the most important was the invention by Frank Gray of a method of transmitting two television signals within a frequency band normally sufficient only for one such signal. The method was based on the utilization of the interstices discovered by Gray (and independently by P. Mertz) between the spectrum lines of a television signal, these lines being spaced at intervals equal to the line scanning frequency. This fact was first disclosed in a memorandum by Hefele and Morrison dated 21 January 1928[89] and was the basis of a patent,[90] in Gray's name, filed on 30 April 1929. Later Gray and Mertz collaborated on the preparation of a paper which is now regarded as a 'classic' in the television literature.[91]

Gray's invention did not have an immediate application in 1929. More than two decades later, however, the great value of his disclosure was appreciated when it became necessary to transmit, simultaneously, the luminance and chrominance signals of a colour television system. For this work Gray received, in 1953, some months after his retirement from Bell Laboratories, the Vladimir K. Zworykin Award of the Institute of Radio Engineers.[92] Normally the award is given to the person who has made the most important contribution to television in the year immediately preceding the award. The justification for granting it to Gray was that it was only during the early 1950s that the industry realized the importance of the interleafing principle in designing the NTSC colour television system.

CONCLUSIONS

Bell Telephone Laboratories commenced their television work at an opportune time, for by 1925 wireless broadcasting was firmly established, the basic components for a rudimentary television system were available to experimenters, interest in television was growing and the Laboratories had just completed the development of still picture transmission apparatus. Ives's choice of Nipkow disc scanners, photoemissive cells and neon glow discharge lamps was a wise one and enabled success to be achieved after a relatively short period of experimentation. Although many different scanning arrangements had been advanced, and were subsequently to be advanced, only the Nipkow disc endured until the late 1930s. Baird had made a similar selection and both he and Ives's group were able, comparatively easily, to adapt their television schemes to demonstrate daylight television, colour television, two-way television, large-screen television and multichannel television. As a consequence the steps which had to be undertaken to evolve and develop these systems had a certain correspondence. No other single inventor-engineer or group of inventor-engineers, anywhere in the world, could exhibit television equipment, during the 1920s, that was inherently so versatile as that of Bell Telephone Laboratories or Baird.

The apparatus of Ives's group was markedly superior in quality of design and construction to that of Baird. The group's demonstrations represented the best that could be given in the 1920s and established standards from which others could judge the worth of their own efforts.

Elsewhere a few inventors and groups were pursuing the solution of the television problem by all-electronic means. The endeavours of Farnsworth and Zworykin in the 1920s and 1930s, and of RCA and EMI in the 1930s, were outstanding. Both Ives and Baird were, until 1933, unimpressed by the possibilities which electronic scanning offered. Their early successes with mechanical scanning equipment and the concurrent difficulties associated with the image dissector and oscillite of Farnsworth, and the iconoscope and kinescope of Zworykin, conceivably may have influenced their thoughts on the way forward in television.

Essentially Ives and Baird delayed for too long the inevitable—with hindsight—move towards high-definition television and the utilization of cathode ray tubes in cameras and receivers, and as a consequence Bell

Laboratories and Baird Television Ltd were overtaken by RCA and EMI. Yet when, in 1933, Ives's group began work on electronic television, the problems identified with it were tackled with enthusiasm and skill. Gray in particular was an inventive and resourceful member of Ives's team. He did not seem to share Ives's entrenched conception of television by mechanical scanning and in 1930 had submitted a comprehensive memorandum on electronic camera tubes. This contained many ideas which could have been investigated. Given that the Laboratories could demonstrate a form of camera tube only two years after the initiation of its electronic television project, it is not improbable that, if Gray's notions had been subjected to experimental scrutiny, Ives and his colleagues could have exhibited a camera tube contemporaneously with Zworykin's disclosure of the iconoscope in 1933.

By contrast, Shoenberg, as Director of Research at EMI, held a firm belief in the ineluctability of the move towards high-definition television by all-electronic means. He formed a powerful team of scientists, engineers and technicians (notwithstanding the depressed business conditions), and worked vigorously in directing his group towards its objective. In less than four years he was in a position to offer the Television Advisory Committee a standard of television which had never been propounded previously and which was to remain the standard for many years.

Zworykin's and Engstrom's teams at RCA were well supported by Sarnoff, the President of RCA. He was far-sighted, optimistic by nature, and never doubted the inevitability of a system of television which would produce, in the average person's home, images of sporting contests, news events and plays. In contrast, Dr Jewett, the President of Bell Telephone Laboratories, had a pessimistic outlook regarding television even after the London television station had been inaugurated in November 1936 and was broadcasting regularly. However, his perspective appears not to have influenced the magnitude of the funds which the Laboratories allocated for television R&D. From 1925 to 1940 these moneys increased as shown below:

1925	$22,900	1931	$69,000	1936	$136,000
1926	24,500	1932	140,000	1937	138,100
1927	36,700	1933	119,000	1938	355,000
1928	86,800	1934	130,000	1939	300,000
1929	80,400	1935	134,400	1940	350,000
1930	56,800				

In 1927 Jewett gave Ives a clear mandate to pursue such fundamental work on television as would safeguard the company's position and advance the art along lines that were likely to be of interest to the company.

Ives felt, in 1931, that a restraint had been imposed on forward planning by the terms of the AT&T–RCA contract. On the other hand the existence of the contract did not inhibit the commencement, in 1933, of the Laboratories' work on all-electronic television. Possibly the lack, in 1931, of a well-defined programme of research and development on television disposed the Laboratories to underrate the importance of their television activities. In that year Ives had written a paper on the problems of television, which was gloomy about the future. Moreover, Ives's suggestions for future work did not embrace electronic television and were all posited on extending his group's basic mechanical scanning scheme. There was a lack of new ideas in his plans.

Acknowledgements

The author acknowledges, with gratitude, the magnanimity of AT&T Bell Laboratories, Murray Hill and Warren, in generously making available to him the many memoranda, reports and other historic documents which formed the primary source material for this paper. This material has been used with the permission of AT&T Bell Laboratories.

Notes and References

1. Russell W. Burns, *British Television, the Formative Years*, (London, 1986), 8–48.

2. Russell W. Burns, 'Early Admiralty interest in television', *IEE Conference Publication of 11th IEE Weekend Meeting on the History of Electrical Engineering*, 1983, 1–17.

3. Burns, op. cit. (1).

4. Russell W. Burns, 'Seeing by electricity', *IEE Proc. A*, 1985, 133: 27–37.

5. R.A. Heising, 'Alexander McLean Nicolson', *Bell Laboratories Record*, May 1950, 221. Anon, 'A.M'L. Nicolson, video pioneer, 69', *New York Times*, 4 February 1950.

6. Alexander McL. Nicolson, 'Television transmission and reception', a memorandum and report to E.H. Colpitts, 13 October 1916, 1–28, Correspondence Folder 200670, Vol. B (1/1/16 to 11/29/16), AT&T Archives, Warren, New Jersey.

7. Alexander McL. Nicolson, 'Television', US Patent No. 1,470,696, 7 December 1917. British Patents Nos 228,961 and 230,401, 7 September 1923.

8. M. Dieckman and G. Glage, 'Verfahren zur Übertragung von Schriftzeichen und Strichzeichnungen unter Benützung der Kathodenstrahlröhre', German Patent No. 190,102, 12 September 1906.

9. B. Rosing, 'New or improved methods of electrically transmitting to a distance real optical images and apparatus therefor', British Patent No. 27,570, 13 December 1907.

10. A.A. Campbell Swinton, 'Presidential address', *Journal of the Roentgen Society*, 1912, 8: 1–5.

11. F. Skaupy, 'Braunsche Röhre mit Gluhkathode, insbesondere für die Zwecke der Elektrischen Bildübertagung', German Patent No. 349,838, 28 November 1919.

12. Vladimir K. Zworykin, 'Television system', US Patent No. 1,691,324, 13 July 1925.

13. George K. Thompson, 'Telephotography', memorandum for file, January 1884, Case 37014, Correspondence Folder 459 (2/18/86 to 9/30/97), Boston File, Radiophony, AT&T Archives, Warren, New Jersey.

14. P.M., 'Television', memorandum, 6 September 1921, Television Folder, File 6.035.3, 8. AT&T Archives, Warren, New Jersey.

15. Burns, op. cit. (4).

16. 'Remarks by Frank B. Jewett at the television demonstration', *Bell Laboratory Record*, May 1927, 298.

17. Herbert E. Ives, 'Television: 20th anniversary', *Bell Laboratory Record*, Vol. 25, May 1947, 190–3.

18. Herbert E. Ives, 'Television', memorandum to H.D. Arnold, 23 January 1925, Case File 33089, Vol. A, 8. AT&T Archives, Warren, New Jersey.

19. Paul B. Findley, 'Biography of Herbert E. Ives', internal report, date unknown, 1-12. AT&T Archives, Warren, New Jersey.

20. Burns, op. cit. (1).

21. J.G. Roberts, 'Invention of Mr. H.E. Ives—system of television, memorandum for file, 14 May 1925, Case File 33089, Vol. A, 1. AT&T Archives, Warren, New Jersey.

22. Ives, op. cit. (17).

23. Anon., 'Television—a group achievement', *Bell Laboratory Record*, May 1927, 316-23.

24. Herbert E. Ives, 'Television', *The Bell System Technical Journal*, October 1927, 551-9.

25. F. Gray, J.W. Horton and R.C. Mathes, 'The production and utilization of television signals', *The Bell System Technical Journal*, October 1927, 560-81.

26. Burns, op. cit. (2).

27. E.L. Nelson, 'Radio transmission for television', *The Bell System Technical Journal*, October 1927, 633-53.

28. Ives, op. cit. (17).

29. Anon., 'Far-off speakers seen as well as heard here in a test of television', *New York Times*, 8 April 1927, 1.

30. Colonel A.S. Angwin, memorandum to the Secretary of the GPO, Minute 51/1929, file 9, 7 June 1928, Post Office Archives, London.

31. A. Dauvillier, 'La télévision électrique', *Revue générale de l'Électricité*, 7 January 1928, 5-23.

32. Herbert E. Ives, 'Development program for television', a memorandum to H.D. Arnold, 4 May 1927, Case File 33089, Vol. A, 1-9, AT&T Archives, Warren, New Jersey.

33. H.S.R., 'Television', memorandum for file, 18 March 1966, Case Book No. 1538, Case File 20348 and Case File 19350, 1-12. AT&T Archives, Warren, New Jersey.

34. Frank Gray and Herbert E. Ives, 'Optical conditions for direct scanning in television', *Journal of the Optical Society of America*, 1928, 17: 423-34.

35. Anon., 'Television shows panoramic scene carried by sunlight', *New York Times*, 13 July 1928, 4.

36. Herbert E. Ives, memorandum to H.D. Arnold, 18 February 1928, Case File 33089, 1-5. AT&T Archives, Warren, New Jersey.

37. Ralph V.L. Hartley and Herbert E. Ives, 'Improvements in or relating to television', British Patent No. 297,078, application date (UK) 19 March 1928 (Convention date, USA, 14 September 1927), issued 19 June 1929.

38. Ives, op. cit. (18).

39. Herbert E. Ives, 'Television', memorandum for file, 10 July 1925, Case File 33089, Vol. A, 1-2. AT&T Archives, Warren, New Jersey.

40. Burns, op. cit. (1).

41. Frank Gray, 'The projection of television images', memorandum for file, 13 February 1929, Case File 33089, 1-5. AT&T Archives, Warren, New Jersey.

42. Frank Gray, 'Recording television images on movie film at television speeds', memorandum for file, 11 February 1929, Case File 33089, 1-4. AT&T Archives, Warren, New Jersey.

43. Herbert E. Ives, 'Improvements in the diffraction of colour photography', *Physical Review*, 1906, 22: 339.

44. Herbert E. Ives, 'Three colour interference pictures', *Physical Review*, 1907, 24: 103.

45. Anon., 'Television in color shown first time', *New York Times*, 28 June 1929, 25.

46. Herbert E. Ives, 'Television in color', *Bell Laboratory Record*, July 1929, 7: 439–44.

47. Anon., 'Television in colour successfully shown', *Telephony*, 6 July 1929, 97: 23–5.

48. Anon., '2-way television in phoning tested', *New York Times*, 4 April 1930.

49. Herbert E. Ives, Frank Gray and M.W. Baldwin, 'Image transmission system for two-way television', *The Bell System Technical Journal*, 1930, 9: 448–69.

50. Herbert E. Ives, 'Some optical features in two-way television', *Journal of the Optical Society of America*, 1931, 21: 101–8.

51. W.C.F. Farnell, memorandum to John Mills, 23 June 1930, Ref. 592-C-WCFF-EO, 1. AT&T Archives, Warren, New Jersey.

52. W.D. Sargent, letter to L.S. O'Roark, 9 July 1931, Ref. 0603-2, 1. AT&T Archives, Warren, New Jersey.

53. W.J. O'C., memorandum to A.W. Page, 23 March 1931, Ref. 0603-2, 1. AT&T Archives, Warren, New Jersey.

54. Herbert E. Ives, 'A multi-channel television apparatus', *Journal of the Optical Society of America*, 1931, 21: 8–19.

55. P.M., op. cit. (14).

56. Burns, op. cit. (1).

57. Swinton, op. cit. (10).

58. George Everson, *The Story of Television, the Life of Philo T. Farnsworth*, (New York, 1949).

59. Burns, op. cit. (1).

60. Herbert E. Ives, 'Future program for television research and development', memorandum for file, 18 May 1931, Case File 33089, 1–11. AT&T Archives, Warren, New Jersey.

61. W.R. MacLaurin, *Invention and Innovation in the Radio Industry*, (New York, 1949), 105.

62. Lloyd Espenschied, 'Announcement of television developments in the AT&T Co. Annual Reports', memorandum, 19 February 1954, 1–3. AT&T Archives, Warren, New Jersey.

63. Lloyd Espenschied, 'When television was in eclipse in the Bell System—Case 37014', memorandum for file, 18 February 1954, 1–3. AT&T Archives, Warren, New Jersey.

64. Anon., 'System of television, Case 33,089, work authorization estimated costs', from file on 'Television and radio history—picture transmission', AT&T Archives, Warren, New Jersey.

65. Frank Gray, 'The cathode ray tube as a television receiver', memorandum to H.E. Ives, 16 November 1926, Case File 33089, 1–2. AT&T Archives, Warren, New Jersey.

66. L.H. Bedford and O.S. Puckle, 'A velocity modulation television system', *Journal of the IEE*, 1935, 75: 91.

67. Herbert E. Ives, memorandum to H.P. Charlesworth, 16 December 1929, Case File 33089, 1. AT&T Archives, Warren, New Jersey.

68. Ives, op. cit. (58).

69. Burns, op. cit. (1), Chapter 18, 423–41.

70. Burns, op. cit. (1), 235.

71. Frank Gray, 'Note on Zworykin's iconoscope', 6 July 1933, Case File 33089, 1-3. AT&T Archives, Warren, New Jersey.

72. O.E. Buckley and Herbert E. Ives, memorandum to H.D. Arnold, 6 July

1933. Case File 33089, 1–2. AT&T Archives, Warren, New Jersey.
73. Herbert E. Ives, memorandum to O.E. Buckley, 20 July 1933, Case File 33089, 1. AT&T Archives, Warren, New Jersey.
74. Frank Gray, 'A suggested outline for development work on a cathode ray transmitter', memorandum for file, 31 July 1933, Case File 33089-1, 1–3. AT&T Archives, Warren, New Jersey.
75. Frank Gray, 'Proposed television transmitters', memorandum for file, 20 May 1930, Case File 33089, 1–21. AT&T Archives, Warren, New Jersey.
76. Frank Gray, 'A proposed cathode ray transmitter', memorandum for file, 3 January 1934, Case File 33089, 1. AT&T Archives, Warren, New Jersey.
77. Frank Gray, 'Projection of images from a cathode ray tube', memorandum for file, 15 January 1934, Case File 33089-1, 1–5. AT&T Archives, Warren, New Jersey.
78. Gordon K. Teal, 'A new photoelectric emitter suitable for use in the iconoscope', memorandum for file, 25 April 1934, Case File 33089, 1–2; 'The potassium-potassium oxide-silver matrix as the photo-sensitive element of the iconoscope', memorandum for file, 14 May 1934, Case File 33089, 1–2. AT&T Archives, Warren, New Jersey.
79. 'Notes of a Conference held in Dr Buckley's office on large image schemes for theatre showing', Bell Telephone Laboratories, 20 June 1934, Case File 33089, 1–8. AT&T Archives, Warren, New Jersey.
80. Herbert E. Ives, 'Iconoscope for transmission from motion picture film', memorandum for file, 6 July 1934, Case File 33089, 1. AT&T Archives, Warren, New Jersey.
81. John R. Hefele, 'Television transmission system using cathode ray tubes', memorandum for file, 31 July 1931, Case File 33089, 1–4. AT&T Archives, Warren, New Jersey.
82. Foster C. Nix, 'Photoconducting television transmitter', memorandum to O.E. Buckley, H.E. Ives, E.F. Kingsbury, M.E. Streiby, J.B. Johnson, C.J. Davison, 6 August 1935, Case File 33089, 1. AT&T Archives, Warren, New Jersey.
83. Burns, op. cit. (1).
84. Anon., 'Observations on Dr F.B. Jewett's testimony before the FCC concerning the division of expenses of the laboratories as between AT&T and Western, and as between local and toll', memorandum for file, 20 July 1938, 1–2. AT&T Archives, Warren, New Jersey.
85. Anon., 'Bell Labs test coaxial cable', *Electronics*, December 1937, 18–19.
86. Ibid.
87. Burns, op. cit. (1).
88. Anon., 'Principal Bell System dates in television', memorandum, November 1946, 1–2. AT&T Archives, Warren, New Jersey.
89. A.G.J., 'Early history of Bell System television work', memorandum for file, 1954, 1–4. AT&T Archives, Warren, New Jersey.
90. Frank Gray, US Patent No. 1,769,920, filed 30 April 1929, issued 8 July 1930.
91. P. Mertz and F. Gray, 'A theory of scanning and its relationship to the characteristics of the transmitted signal in telephotography and television', *The Bell System Technical Journal*, July 1934, 13: 464–516.
92. A.G.J., op. cit. (89).

Technology Transfer in Russian Electrification, 1870–1925

JONATHAN COOPERSMITH

It is impossible to envision Russian industrialization of the late nineteenth and early twentieth centuries without the large flows of technologies, moneys and people from West to East.[1] Technology transfer took several forms during this half-century, including not only the exportation of manufactured goods and manufacturing technology but also the less visible but extremely important flows of 'stocks of knowledge':[2] people, information and ideas. As the First World War proved, Russia was heavily dependent on European—particularly German—electrotechnology and finance. This dependency developed voluntarily as Russian engineers, managers and financiers saw themselves as part of the larger international community and gravitated towards Germany.

This paper examines the foreign component in the development of Russian electrotechnology from 1870 to 1925. The emphasis is less on materials, equipment, financing and Tsarist industrial and tariff policies, which have been covered elsewhere,[3] than on institutional, personal and conceptual links. My focus is on the importance of the transfer and acceptance of ideas, primarily during the early revolutionary period when the Russian electrical engineering leadership finally had the freedom and the political support to chart the country's future course of electrification. The result was the GOELRO plan of state electrification, which moulded the latest Western technological advances into a wide-ranging programme of economic and social reconstruction.

Although nationalist pressures increased, particularly after 1914, the Russian electrotechnical leadership considered itself part of a larger international community and promoted electrification plans that followed trends in Western electrotechnology despite technological alternatives that were economically and politically appealing. The activities, ideas and interests of Russian inventors in the early decades of the electrical industry paced those of their Western counterparts. In development, diffusion and application, however, the advantages lay with the more hospitable economic and social environment of the West, with its larger, more advanced technical and financial base. Indeed, Russian inventors working in the West were the major exception to the one-way flow of technology from West to East.

It is important to realize that Russian advocates of electrification were active, not passive, recipients of foreign electrotechnology who employed foreign activities and ideas to legitimate and justify their particular courses of development. Yet Russian electrical engineers suffered from an almost schizoid split between the desire for the latest Western technology and the desire to develop an indigenous technological base. Although efforts to Russify the electrotechnical industry forcibly date back to the mid-1880s,[4] the state did not make the development of an independent indigenous capability a high priority until 1925.[5]

Technology was transferred via firms, government agencies, professional organizations, the media and individuals. Foreign firms exported technology to Russia via subsidiaries, direct sales and cooperative agreements. The main Russian actors were electrical engineers and utility administrators, but government agencies and, later, the Communist Party also played major roles. The Ministry of Finance and the Ministry of Trade and Industry strove to create an institutional infrastructure and political climate conducive to industrial development, mainly after 1890.[6]

The Russian army and navy served as the major conduit of foreign technology from the early 1870s to the late 1880s because of their promotion of electric lighting for military purposes.[7] The army and navy introduced new technologies into Russia, such as the Alliance generator in 1869 and the Jablochkov arc light in 1878, while also promoting domestic manufactures.[8] The initial introduction and diffusion of electrotechnology into Russia by the army and navy can be viewed as the military sector keeping up with Western activity and the civilian sector lagging behind, a frequent pattern in Russian history.

EQUIPMENT AND FINANCE

German firms were the major foreign actors in Russia, reflecting both German leadership in electrotechnology and an aggressive, export-oriented industry.[9] Determined, persistent and thorough marketing fortified the technological strengths of German firms. The German businessman in Russia spoke Russian, carried brochures and catalogues written in Russian, and could arrange long-term credit.[10] German equipment dominated Russian electrotechnical imports. In 1906–10, a period of rapid utility and industry expansion, German imports comprised 80–90% of all imports by weight.[11]

The leading German firms, Siemens & Halske and AEG, competed with Brown-Boveri, Westinghouse, Metropolitan Vickers, other foreign firms, and Russian firms. Simple 'low technology' products, such as cables and small electric motors, initially were imported until a domestic manufacturing base developed. More complex 'high technology' equipment, such as steam turbines, remained available only abroad. None the less, the domestic electrotechnical industry, including foreign subsidiaries, grew to supply half of the country's equipment by 1914.[12] The First World War, which sharply curtailed German imports, revealed that Russia's dependence on foreign equipment ranged from turbo-generators and transformers to instruments and incandescent lights.[13]

Siemens & Halske had entered Russia in 1853 to construct telegraph lines for the government. The St Petersburg-based firm gradually expanded into electrical manufacturing, copper mining and other industrial activities.[14] In July 1886, the Ministry of Finance gave Siemens & Halske the right to form the separate Company for Electric Lighting (Obshchestvo Elektricheskogo Osveshcheniia), commonly known as the 1886 Company.[15] German stockholders, including the company's president, Karl Siemens, the brother in charge of Russian operations, predominated. The 1886 Company operated Russia's first successful utility in St Petersburg and expanded to Moscow in 1887 and Łódź in 1908. The 1886 Company thus controlled three of the empire's four largest utilities in major cities with rapid growth, which not only produced dividends of 6–10% but also necessitated constant reinvestment, new stock offerings and new equipment.[16]

German-owned concessions operated Russia's most technically advanced utilities. In 1897, the 1886 Company's Georgievsk station in Moscow became the first Russian utility to use three-phase alternating current, a more efficient form of transmitting electricity than direct current.[17] The Marxist engineer Robert E. Klasson, who had worked under Mikhail O. Dolivo-Dobrovolskii in Germany and constructed the first large Russian three-phase a.c. station at the army's Okhtensk gunpowder factory in 1895, directed the project.[18] In 1913–14, Klasson would construct Russia's first regional station, Elektroperedacha, with the aid of German capital. The Georgievsk station served as the keystone of the Moscow section for two decades, evolving from vertical compound engines in 1897 to turbines in 1902. The station played an 'exceptionally important role' in the development of Russian electrotechnology by training Russian engineers and technicians and developing and diffusing new equipment and methods.[19]

Two other foreign firms, the German-owned Helios and the Belgium Company for Electric Lighting, operated utility concessions in St Petersburg after 1900.[20] German-based concessionaires also provided electricity in other cities of more than 250,000 people, such as Kiev and Baku, in a demonstration of superior financing, technical expertise and experience. These foreign-owned utilities introduced new equipment and techniques into Russia, trained administrative and technical staff, and pioneered the industrial usage of electrical energy.

Utilities used a mix of foreign and domestic equipment in a blend that reflected economic and technical demands. The larger the utility, the more likely it was to use high-technology imported turbo-generators. Smaller utilities tended to use domestically manufactured vertical or horizontal engines.[21] The economics of Russian operations demanded equipment that often was either highly specialized or very primitive. After 1891, St Petersburg utilities used boiler equipment designed for Cardiff smokeless coal.[22] In a reflection of the high cost of capital, utilities in the capital employed hundreds of workers instead of automated equipment to move coal from barge to power-plant.[23]

Trams were one area where German firms did not dominate. As in other countries, Belgian firms dominated the tram market in Russia because they offered technically mature systems accompanied by managerial expertise and financial backing.[24] In 1892 a Belgian firm opened Russia's first electric traction tramway in Kiev, a city where steep hills made horse trams impractical in some sections.[25] By 1913, Belgian concessions operated twenty electric trams, i.e. nearly half the country's forty-one electric trams, plus two horse trams and one steam tram.[26] The Brussels stock market listed 188 tram companies in 1913, including nineteen Russian companies; by contrast, neither the St Petersburg nor the Moscow stock market listed a single tram company in 1913.[27] Belgian-based capital, which included funds from other countries, constituted two-thirds (137 million roubles) of the 206 million roubles invested in trams between 1909 and 1914.[28] This Belgian domination continued that country's international entrepreneurial involvement in horse trams.

Financing is the underlying *sine qua non* of commercial technologies. The best equipment in the world is useless without the means to purchase and operate it. Although the exact numbers remain a source of contention, foreign investment accounted for significant amounts of Russian government and non-government capital formation.[29] Electrotechnology proved no exception: foreign banks and companies financed the bulk of prewar electrification, usually working with a Russian bank.[30] Of 270 million roubles invested in utilities by 1914, German money accounted for nearly half and Russian funding for less than 10%; other countries provided the rest.[31] German firms provided 65% of the 120 million roubles foreigners invested in domestic manufacturing.[32]

Foreign financing permitted Russian electrification to develop as quickly as it did because the Russian credit market could not supply the capital needed to construct utilities. Tsarist restrictions on the Russian stock exchange hindered the efficient creation and transfer of capital and thereby increased the country's dependence on foreign capital to finance the development of capital-intensive industries, such as electrification.[33] Beyond the obvious delays in utility expansion and diffusion, domestic limits on credit meant that Russian utilities never challenged and transformed the country's financial markets as utilities did in the West.[34] Consequently, foreign financing both delayed the formation of capital markets and accelerated the diffusion of electrotechnology in Russia.

Technology transfer played a vital role in equipping and financing Russian utilities. The flow of foreign ideas and information proved equally important in shaping the direction of electrification in Russia.

INDIVIDUALS, INSTITUTIONS AND PROFESSIONAL LINKS

Perhaps the least visible and most important component of technology transfer is the flow of people. During the half-century of Tsarist industrialization, tens of thousands of Russians went abroad for technical and scientific training.[35] Electrical engineers proved no exception. Of a sample of forty prominent pre-revolutionary electrical engineers, two-thirds studied or worked abroad at some time in their careers.[36] They travelled abroad for further education, to work, to attend international congresses and exhibitions, and to tour industrial installations. Trips abroad enabled them to meet and talk with their Western counterparts and to see and work on the latest technical developments. These electrical engineers linked Russia to the international electrotechnical community.

Individuals proved the major exception to the one-way transfer of technology. Russian engineers and scientists worked in Europe and the United States, where they contributed greatly to the development of electrotechnology. The four most prominent examples were Aleksandr N. Lodygin, Pavel N. Jablochkov, Mikhail O. Dolivo-Dobrovolskii and Achilles de Khotinskii. Only the first did his most notable work in Russia; the others achieved international recognition because of their research and work in the West.

Lodygin invented an incandescent electric light in the early 1870s, but failed commercially and spent most of his professional life after 1880 in France and the United States until he died in 1923. Jablochkov developed the world's first commercially successful arc light in Paris in 1875.[37] Working for the German firm AEG, Dolivo-Dobrovolskii in 1891 transmitted three-phase alternating current 170 kilometres from Lauffen to an electrical exhibition in Frankfurt, a major milestone in the development of technically and economically feasible long-distance transmission of electric energy.[38] Achilles de Khotinskii, a former naval officer, manufactured light-bulbs in Russia and the West in the 1880s.[39] Both Dolivo-Dobrovolskii and de Khotinskii initially left Russia involuntarily because of their political activities. The inability of inventors to succeed in Russia marked not personal inadequacies but more general handicaps of modernizing the country.

Although a major conduit of information about Western electrotechnology, engineers abroad comprised a fraction of the Russian electrotechnical community. The majority received information mainly from foreign and Russian periodicals. A 1913 survey of graduates from the élite St Petersburg Polytechnical Institute found that they read thirteen electrical journals in all. Three were Russian, three English or American, and seven German. Half of the respondents (49%) read the German *Elektrotechnische Zeitschrift*, compared with 70% who read *Elektrichestvo*, the main Russian electrical journal.[40]

Russian electrotechnical periodicals contained numerous translated foreign articles, Russian articles on Western developments and sections devoted to foreign activities. When *Elektrichestvo* began in 1880, it had a table of contents in Russian and French. By the late 1880s and 1890s, French articles declined in number, and articles of German and British origin increased. American articles did not reach significant numbers until the 1910s.

This shift in coverage corresponded to the shift in the frontiers of electrical engineering from Paris to Berlin, centre of the German electrical industry. The transfer of German knowledge reflected German dominance of the Russian electrical market. The technical language was German.[41] More than 90% of the St Petersburg polytechnic graduates found knowledge of a foreign language indispensable. Fifty-five per cent knew German; only 28 per cent knew English.[42]

Even the first effort by the Electrotechnical Section of the Imperial Russian Technical Society to publish statistics on Russian utilities depended on German information.[43] Russian engineers accepted foreign standards, but only after participating in the international commissions that established these norms.[44] Indeed, during the First World War, when German imports stopped, Russian engineers tried to maintain German standards instead of switching to American standards.[45]

The institutional links between Europe and Russia sometimes proved stronger than intra-Russia ties. There was no Russian association of utilities until 1917, but in 1914 twenty Russian utilities belonged to the Vereinigung Deutscher Elektrizitätewerke, a German association of utilities.[46] The most important foreign technical society for Russian electrical engineers was the German Verein Deutscher Elektrotechniker. In 1888, Russians comprised 54 of the 1,452 VDE members, or about half the active membership of the Electrotechnical Section.[47] Corresponding interest did not exist in non-German electrotechnical societies.[48]

One unsuccessful area of technology transfer was organizational structures. Several proposed Russian organizations for electrification featured Western components. Just before the February 1917 revolution, the War Ministry proposed a hydrostation for Petrograd with management by a special committee with powers 'like government committees in England and the United States'.[49] After the February 1917 revolution, the first high-level state organization to promote electrification, the Council for Electrical Affairs (SED) of the Ministry of Trade and Industry, proposed the creation of an inter-departmental committee of experts and representatives of concerned organizations, based on the French model.[50] Early in 1922, a commission of Glavelektro, the state administration for electrotechnology, proposed a state-controlled 'Elektrobank', based on German and Japanese models, to finance all areas of electrification.[51] The attraction of Western organizational models lies as much in the bankruptcy of domestic models as in the attraction of foreign structures, for which no Russian equivalent existed. The failure of these proposals lay not in their foreign component but a lack of political support.

The case can be made that Russian electrical engineers had no choice but to submit before the flood of foreign technology and imports. The same cannot be said about the willingness with which these engineers adopted technocratic Western ideas of large-scale electrification.

IDEAS

Western ideas greatly influenced the political and technological evolution of Russian electrification. Foreign data and concepts such as regional stations and the use of electric light and power for political, economic and social goals provided evidence and legitimacy for Russian progressives and reformers. As the discussion on the future course of electrification will show, advocates of both decentralized and centralized electrification used Western concepts to advance their cause, but the promoters of regional stations mustered both the weight of Western experience and greater domestic political support.

In Russia, the decade after the 1905–6 revolution saw a growing number of proposals by engineers and others to use electric energy 'as a mighty factor in contemporary social-economic development'.[52] Foremost were assisting small-scale industries to compete with big business; the 'democratization' of lighting; spreading electric light and power to rural areas; and municipal control of utilities. Anti-foreign and pro-Russian industry elements intertwined with several of these proposals, which formed part of a larger technocratic movement among Russian engineers and a progressive movement in the municipalities.[53] By 1914, these proposals had grown into schemes for widespread, regional electrification.

Significantly—and despite the pro-Russian and anti-foreign elements of some proposals—most of these ideas and the supporting data originated in the West.[54] Russian efforts at municipalization and other progressive efforts followed Western 'good government' drives that coupled local control and ownership with efficient operations.[55] Russian and Western proposals shared other characteristics: ideas somewhat ahead of what was economically and technically feasible, and a major influence in the wartime and post-war debates over electrification.

The industrial demands for electricity during the First World War increased the political influence of the electrical engineering community, which advanced the concept of large-scale electrification based on regional stations. Foreign concepts and data provided the foundation for these ideas, including a November 1915 proposal by Marxist engineer Gleb M. Krzhizhanovskii, the commercial director of Elektroperedacha, the country's only pre-revolutionary regional station, to supply the country's Central Industrial Region with only eight peat-fired stations.[56] He based his position on contemporary American criteria.

The war also focused thinking on how to strengthen Russian industry and overcome its obvious dependence on German technology.[57] Some, such as Vladimir I. Kovalevskii, director of the Department of Trade and Industry from 1892 to 1900 and president of the Imperial Russian Technical Society from 1906 to 1916, almost welcomed the war as an opportunity to throw off the shackles of German 'economic oppression' and develop into an economic power capable of standing up to foreign capitalism.[58]

The idea of large-scale state electrification advanced quickly as Russia moved into a destructive period of revolution and civil war. By October 1917, key élites in government, industry and the engineering communities had experienced first-hand some form of state planning and control and expected more involvement in the immediate future.[59] The October Revolution added a mix of Utopian visions and revolutionary dreams that further strengthened interest in employing new technologies and planning to change the economic and social foundations of the country.[60]

In December 1917, the government formed the Electrotechnical Section, the predecessor to Glavelektro, to handle 'the production of electrical energy'[61] and the underlying manufacturing base.[62] The announcement of the section's creation proclaimed that only the creation of a network of state-owned regional stations

> can place the Russian economy on the level demanded by the international situation. . . . For the reconstruction of the national economy after the end of the war, the first question is about receiving inexpensive energy by the directed and the planned [construction] of regional electric stations of high voltage (120,000 V) from 'white' (waterfalls), 'grey' (peat) and black coal. . . . These plans on a State scale *already have been worked out (Germany) or are being worked out in all countries*.[63]

Knowledge of foreign activities and ideas, particularly the wartime German experience with state planning, contributed to the growing Russian advocacy of state capitalism.[64] A major German influence on Russian thinking was Karl Ballod, a professor of economics in Berlin. His *Der Zukunftstaat* (The Future State) showed how to organize a centrally planned socialist economy. The first edition, published in Germany in 1898, appeared in several Russian translations from 1903 to 1906. The second edition, which appeared in Germany in 1919 and in Russia in 1920, helped convince Lenin and others that a planned economy was feasible.[65] Ballod's ideas influenced Russians as early as 1898, when Aleksandr I. Ugrimov, a GOELRO specialist on agriculture, heard him lecture at Leipzig.[66]

While Karl Ballod influenced Russian concepts of planning, another German, Georg Klingenberg, influenced Russian concepts of regional power stations.[67] Klingenberg, the head of power plant design for AEG, pioneered the concept of the regional electric station in Germany, site of several efforts in prewar regional electrification and utilization of low-quality fuel.[68]

German experience and theory were important because they gave Russians, Marxist and non-Marxist alike, the justifying legitimacy of foreign involvement as well as guidelines on what to do. Equally importantly, the revolutionary destruction of the Tsarist political base and Utopian goals gave Russian electrical engineers the opportunity in 1920 to surpass the West in introducing Western concepts of large-scale electrification into Russian society.

THE GOELRO PLAN OF STATE ELECTRIFICATION

In December 1920, the State Commission for the Electrification of Russia (GOELRO) submitted a far-ranging plan for the electrification and industrialization of Soviet Russia. Approved by the state and Communist Party, the GOELRO plan made electrification a state technology under Lenin's slogan, 'Communism is Soviet power plus the electrification of the whole country'.[69]

GOELRO had a choice of three technological paths for future electrification: a radical-conservative path of building very large regional stations, a conservative path of expanding existing utilities, and a radical-radical path of rapidly electrifying the countryside to bring the peasantry into the civilized world of the town-dweller. Each choice had different technical implications. Regional stations and long-distance transmission networks required substantial funding and a greater reliance on foreign technology than the alternatives, which did not require the high technologies available only in the West. Indeed, the radical-radical option implied support of decentralized industries, particularly the traditional *kustar* (handicrafts) industries, to build small electric stations. In 1920, GOELRO chose the radical-conservative path which centralized decision-making and concentrated resources in regional stations.

Western influence ran deeply throughout the entire GOELRO plan and the accompanying discussions. Foremost was Western progress in electrical engineering, which provided inspiration and justification: 'If they can do it, so can we—and since they are doing it, we should.' Furthermore, Russian electrical and political leaders expected Western technology and capital to play a large role in implementing the GOELRO plan just as they had in Tsarist times.

In contrast to the preface of the 1955 reprint, which claimed that 'independence and defensive capacity of the Soviet state were the central, leading goals' of electrification,[70] the GOELRO plan assumed a productive economic interdependence between the capitalist countries and the world's premier socialist state. Western technology would allow Russia to overcome its wartime devastation, while 'It is vital to display before Europe the possibility of using the riches of West Siberia, especially in agriculture, for an escape from the forthcoming world food crisis.'[71] In agriculture, Boris I. Ugrimov, the GOELRO agricultural expert, cited wheat exports as the reason for feeding Soviet citizens at a level just above hunger.[72] The Gorev–Shvartz plan for northern electrification envisaged the development of an export-oriented aluminium industry aided by the transforming of Murmansk into a deep-sea port with concomitant development of shipbuilding and a transportation network.[73] The Dniepr river development envisioned Aleksandrovsk as a second Odessa linking southern Russia to Europe.[74]

Underlying the wide-ranging GOELRO plan was the assumption of normal international economic relations. The export of revolution and international proletarian solidarity might be publicly espoused, but conventional trade dealings would be the order of the day.[75] This rather astounding presumption—astounding in 1920 when the young state was just re-establishing trade links and remained diplomatically unrecognized by the major powers—owed much to the world view of the Russian electrical engineers. The GOELRO planners, Marxist and non-Marxist, were electrical engineers who had lived in the West, worked for foreign firms and followed Western developments. They wholeheartedly advocated the continued flow of advanced technologies into Russia from abroad. The question was not 'if' but 'how much'. If the country was to be electrified, the job should be done properly—which meant extensive cooperation with the West.

This does not sound terribly shocking, but in 1920 it was. GOELRO viewed the West not as the site for the next proletarian revolution but as a source of technology and financing. Under GOELRO's plan, Soviet Russia would not build socialism autarkically in one country or export the dictatorship of the proletariat but resume international relations, albeit on its own terms. The Russian electrical engineering leadership very much wanted to return to the international electrotechnical community, where they thought—correctly—they belonged. GOELRO was not just a plan for reconstructing Russia, but also a statement that Russian electrical engineers were members of a larger engineering fraternity as well as the technocratic modernizers for the new state. It was a statement that other Russian scientists and engineers would echo.[76]

The Eighth All-Russian Electrotechnical Conference, held in October 1921 to demonstrate engineering support for the GOELRO plan and rebuke those 'sceptics and whisperers who have maliciously mixed the term electrification with the term "electrofiction" [*elektrofiktsiia*]',[77] demonstrated the continued willingness of Russian electrical engineers to take their ideas, as well as their equipment, from the West. Speakers often prefaced their reports by noting, 'here it is necessary to say in advance that the equipment must be received from abroad.'[78] GOELRO chairman Krzhizhanovskii claimed that, because of improved access to foreign technical literature since 1920, 'Now we can refer in defence of our position to a whole series of first-class West European authorities.'[79] Who could object to realizing the visions of a Steinmetz or Ostwald? The copious information presented about Western activities allowed Krzhizhanovskii to claim that 'these data graphically show that we were right when we established [GOELRO, for] all progress in worldwide technology is tightly linked to electrification.'[80]

Both advocates of GOELRO's centralized regional stations and proponents of small-scale rapid rural electrification used Western data and ideas to support their platforms. The issue boiled down to resources and decision-making. Should electrification be directed from above or arise democratically from below?[81] Was local electrification 'economically illiterate', wasteful and, as viewed in the West, neither economically nor technically rational?[82] Or were small stations the best means to bring electricity to the rural regions?[83] City officials opposed not electrification, which would transform the country 'into a second North America', but central control by the state electrotechnical administration, Glavelektro, which 'terrorized' local authorities by depriving them of materials and equipment.[84]

Advocates of small-scale electrification claimed it could transform the countryside years before GOELRO's industrially oriented regional stations would reach rural areas.[85] Unless it provided this broad-based electrification, the GOELRO plan would follow the old foreign-dominated Tsarist path of large-scale centralized development.[86] Furthermore, small stations would use domestically manufactured equipment and save valuable foreign exchange and revive local and domestic industries.[87]

Not surprisingly, the Eighth All-Russian Electrotechnical Congress resolved that the GOELRO plan of large-scale electrification and industrialization 'on the whole is the correct scheme by which to construct a State planned economy'.[88] For GOELRO's supporters, final justification came from the analogy of how the United States emerged battered from its Civil War only to host the world's most impressive technical exhibition a decade later.[89] Even in the rise from destruction, the analogy was foreign.

The GOELRO plan remained in the tradition of Western development and Bolshevik centralization. If the Congress had fully supported the promise of small-scale electrification, that would have been the true revolution. Instead of building on Western lines of development, a uniquely Soviet decentralized model of electrification could have emerged. The social effects of electricity, especially in the rural areas, might have been as revolutionary as GOELRO intended. This decentralized approach never captured the support of the Party or the electrical engineering leadership, both of whose interests favoured centralized control. The road not taken demanded different drivers from the GOELRO engineers, whose professional training and experience promoted centralized networks instead of smaller, isolated independent systems.

In the West as in Russia, the First World War sparked great interest in large-scale engineering projects for social goals and contributed to the post-war popularity of technocratic thinking.[90] Russia was one of several countries where engineers and politicians tried to expand and rationalize electrification along lines of bigness, nationalization and efficiency, but the only country where such plans became state policy.

The three major Western electrification proposals of the 1920s were the unification of the Ruhr, 'Giant Power' of Pennsylvania, and 'Superpower' in the American North-East.[91] Want of sufficient political support and opposition from established economic and political institutions, including railways, utilities and engineering societies, ultimately defeated these radical-conservative proposals, while rural electrification had to wait until the TVA a decade later. Like these Western proposals, the GOELRO plan was not so much a departure from contemporary trends as their directed extension. The theoretical currents that shaped GOELRO were formed in the West, but in technology-fascinated Soviet Russia they became state policy.

Why did Russia, instead of the more industrialized countries, adopt most completely the rationale of the machine age? How did Russia, which took so much from Western ideas, reach—on paper—technocratic goals before the Europeans and Americans? The answer is revolution. The October Revolution brought to power a regime that believed in the promise of machines to liberate.[92] The widespread economic devastation and depreciation of the old regime, combined with the revolutionary hope and expectations for the future, explain why this Utopian plan of 'electrofiction', grounded in Western ideas, became state policy in Soviet Russia and not bourgeois Europe. The strong attraction to GOELRO among non-Bolsheviks and the widespread societal visions of a postwar Russia as a planned, reconstructed, rationalized country show that this interest extended far beyond a small circle of engineers and revolutionaries. The West spawned and nurtured the basic ideas underlying GOELRO but not the intense desire for nor the concomitant political acceptance of technology as a means of modernization.[93]

Implementation did proceed smoothly. GOELRO had assumed, with important caveats about normal economic conditions, that foreign financing and technology would play a major role in electrifying Russia.[94] The flow of Western electrotechnology resumed during the New Economic Policy (NEP), a period of economic relaxation, but neither on the scale desired nor with the prewar degree of foreign control. The framework for exchanges had changed greatly since 1917, when foreign firms owned and operated Russian companies. Now, foreign firms dealt directly with the Commissariat of Foreign Trade and indirectly with state-owned and state-operated trusts and firms. Independent direct investment ceased to exist; Western firms entered into formal agreements for technologies and financial assistance. Soviets, not foreigners, made the decisions. The advisers and critical equipment, however, remained Western.

With the exception of ASEA's 1924 agreement with Elektroselstroi, a Soviet firm promoting rural electrification,[95] neither foreign loans nor manufacturing concessions played a significant role until the First Five Year Plan. Foreign governments proved hesitant to offer recognition and trade privileges. The political difficulties of diplomatic recognition, nationalization of foreign industry, Soviet repudiation of Tsarist foreign debt, concessions less favourable than desired and fear of Bolshevik-fermented revolution made foreigners reluctant to invest.[96] Soviet financial policy, which tried to maximize exports and minimize imports, further restricted access to the West.[97] Communist fear of an alliance between foreign industrialists and Russian peasants, whereby inexpensive foreign goods would undercut the state's industries, further restricted imports.[98]

Although financial and trade links proved disappointing, Russian electrical engineers resumed their participation in the world electrotechnical community, including participating in the International Electrotechnical Commission and the 1924 World Power Conference in London.[99] Western models guided Soviet standardization efforts.[100] Foreign firms sent men and new technologies to Russia. The pages of *Elektrichestvo* soon filled with advertisements from familiar firms: Siemens & Halske, AEG, Metropolitan-Vickers, General Electric, and ASEA.

Imports of equipment resumed, although compared with 1913, postwar imports had a smaller market share and total value. Electrotechnical imports doubled from 1921 to 1926 by value, but their share of the market dropped sharply from 57% in 1921–2 to 20% in 1925–6 as domestic production increased (see Table 1).[101] Regional stations depended on high-technology imports (see Table 2). The decision to pursue regional stations necessitated a continued voluntary reliance on Western technology and construction expertise. Only one of the first seven regional stations, Kizelov, did not import major systems, because it used machinery from the Tsarist Oranienbaum station.[102]

Foreign equipment speeded the reconstruction and expansion of urban stations, but was not essential. Existing utilities sought imports because

Table 1 Electrotechnical sales, 1913–1926 (millions of prewar roubles)

Year	*Domestic*		*Imports*	*Total*
	State	*Local**		
1913	80 (60)		54 (40)	134
1916	96 (71)		39 (29)	135
1921–2	5 (24)	4 (19)	12 (57)	21
1922–3	16 (44)	6 (17)	14 (39)	36
1923–4	40 (63)	9 (14)	15 (23)	64
1924–5	54 (61)	14 (16)	21 (24)	89
1925–6	85 (62)	24 (18)	28 (20)	137
Postwar total	200 (58)	57 (16)	90 (26)	347

* *Kustar* (local handicrafts) industry and other non-state enterprises.

Figures in parentheses are percentages of all sales.

Source: Elektrobank, *Finansirovanie elektrokhoziaistva (dva goda raboty Elektrobanka)*, (Moscow-Leningrad, 1927), 6.

Table 2 Imports for regional stations, 1921–1925

Station	*Equipment*
Red October	10 MW prewar Brown-Boveri turbine Walter boilers with Makarev fireboxes Bruno boilers 10 MW Czech turbo-generator
Nizhegorod	Babcock & Wilcox boilers 10 MW AEG turbo generators
Shterov	10 MW Vickers turbo generators Babcock & Wilcox boilers Combustion Rationelle fireboxes
Kashira	6 MW Brown-Boveri turbo generators Babcock & Wilcox, Garbe, Sterling boilers
Volkhov	12 MW ASEA turbines
Shatura	16 MW Bruno turbo-generators Siemens generators Garbe boilers

Source: Obzor sostoianiia rabot po krupnomu elektrostroitelstvu na 1oe Oktiabria 1925 g, (Moscow, 1926), 19, 22, 24, 26, 29, 30, 33.

they were less expensive, more reliable, more quickly delivered and better built than their domestic equivalents.[103] Only small rural utilities, which relied on *kustar* manufactures to provide the necessary low-technology materials and equipment, did not utilize foreign equipment.[104]

The West, defined increasingly as the United States, continued to serve as a model and moving target as well as a source of supply. What better affirmation of the vigour of Soviet construction could there be than to label it American? At the Volkhov hydrostation, symbol of the new Russia, according to a propagandist, 'Here you see the current America—noise, thunder—all, all America. In a word, there is not a Russian approach but an American tempo.'[105] Yet even as 'Fordism', 'Taylorism' and 'Amerikanizm' captured the enthusiasm of Soviet modernizers,[106] the government began to shift its economic priorities from the technologies of the second to those of the first industrial revolution; that is, from the science-based chemical and electrical technologies to metallurgy and mining.[107] The Party debate in 1924–6 about the future economy moved towards more directed, large-scale industrialization.[108] The Fourteenth Party Congress in December 1925, 'a decisive landmark in the progress of Soviet planning', approved the rapid, planned expansion of industry to provide the economic underpinning for 'socialism in one country'.[109]

Although still a source of emulation, the West increasingly was portrayed as the 'external front' waging an economic war against the Soviet state, with 'Ford and his system the main elements of this struggle against us'.[110] A defiant attitude emerged, which led to assertions that 'energy is the base of one of the most important commanding heights which must be in the hands of the government.'[111] Despite the high costs, the Soviet government and the Communist Party embarked on a course of economic self-sufficiency seen as necessary to preserve and promote 'socialism in one country'. This meant a shift from importing equipment to producing the means of production so that in 'circumstances of capitalist encirclement' the country would not become an adjunct of the world capitalist economy.[112]

After 1925, the Communist Party and not the electrical engineering community set a future course of electrification that differed significantly from the 1920 GOELRO plan. The goal of regional electrification remained the same, but now as a means to industrial self-sufficiency and a state-induced separation from the international electrotechnical community.

CONCLUSION

The massive role of the West as a supplier—of justification, data, ideas, equipment and financing—could, but should not, be viewed simply as evidence of Russia's dependence on technology from abroad. Rather, the Western role shows the extent to which Russian electrical engineers considered themselves part of the international electrotechnical community. Russian electrical engineers were not passive recipients of foreign goods, but filtered and fitted foreign concepts and equipment into a Russian environment. Nor was the transfer all one-way: while the main purpose of setting its subjects abroad to work and study was to benefit the Tsarist state, the international community benefited also, especially from engineers who spent most of their careers in the West.

The GOELRO plan showed the penetration of Western ideas and the political skills of Gleb Krzhizhanovskii and other electrical engineers. Was Russia unusually receptive to foreign ideas? Or did circumstances of revolution and a politically astute electrical engineering leadership account for the creation of the GOELRO plan? The failure of similar large-scale schemes in other countries suggests the uniqueness of the Russian environment, but also the worldwide spread of concepts of state electrification.

The foreign influence on Russian electrification demonstrates the importance of ideas and ideology as well as the more visible forms of technology transfer. Not all ideational influences are as marked as the GOELRO plan, but, as the concentration of Third World researchers on Western-determined priorities and projects instead of more domestically oriented needs indicates, it is a real issue. Hardware rarely transfers without accompanying ideological and ideational components. The history of technology must include these 'invisible' factors.

Acknowledgements

Support for this research came from an International Research and Exchange Board (IREX) fellowship for Moscow State University, an Institute of Electrical and Electronic Engineers Postgraduate Fellowship for time at the MIT Program in Science, Technology and Society and the Harvard Russian Research Center, and a Title VIII fellowship of the Department of State Soviet-Eastern European Research and Training Act of 1983 for time at the Hoover Institution.

Notes and References

1. 'The West' is an ambiguous term, often used as a (higher) standard of comparison with Russia. The West here refers to the more technologically advanced countries of Germany, Great Britain, France, and the United States.

2. Simon Kuznets, *Toward a Theory of Economic Growth*, (New York, 1968), 62.

3. V.S. Diakin, *Germanskie Kapitaly v Rossii*, (Leningrad, 1971); S.A. Gusev, *Razvitie Sovietskoi Elektrotekhnicheskoi Promyshlennosti*, (Moscow, 1964); Guenter S. Holzer, 'German electrical industry in Russia: from economic entrepreneurship to political activism, 1890-1918' (unpublished PhD thesis, University of Nebraska, 1970, AA87108584); Walther Kirchner, 'The industrialization of Russia and the Siemens firms 1853-1890', *Jahrbücher für Geschichte Osteuropas*, 1974, 22: 321-57, 'Siemens and AEG and the electrification of Russia, 1890-1914,' *Jahrbücher für Geschichte Osteuropas*, 1982, 30: 399-428, and *Die Deutsche Industrie und die Industrialisierung Russlands, 1815-1914* (St Katharinen, 1986).

4. 'Otkrytie tekhnicheskogo uchilishcha pochtovo-telegraficheskoi vedomosti', *Elektrichestvo*, 1886, 15: 166-7; 'Khronika', *Elektrotekhnik*, 1897, 2: 159-60.

5. 'Elektrifikatsiia SSSR (rezoliutsiia soveshchanii po elektrifikatsiiu)', *Izvestiia*, 1 July 1925, 5.

6. Theodore H. Von Laue, *Sergei Witte and the Industrialization of Russia*, 2nd edn, (Philadelphia, 1971), 93-9; Hans Rogger, *Russia in the Age of Modernization and Revolution, 1881-1917*, (London, 1983), 102-5.

7. Jonathan C. Coopersmith, 'The role of the military in the electrification of Russia, 1870-1890', in E. Mendelsohn, M.R. Smith and P. Weingart (eds), *Science, Technology and the Military*, 1988, XII: 291-305.

8. 'Otchet predsedatelia uchenogo otdeleniia morskogo tekhnicheskogo komiteta i komiteta morskikh uchenykh zavedenii za 1871', *Morskoi Sbornik*, 1872 (122), 1: 9; Rondolphe van Wetter, *L'Éclairage électrique à la guerre*, (Paris, 1889), 82; Em. Alglave and J. Boulard, *The Electric Light: Its History, Production and Applications*, (New York, 1884), 393; Lev D. Belkind, *Pavel Nikolaevich Jablochkov*, (Moscow, 1962), 168-70,

176–8; N. V-v., 'K voprosy o polozhenii russkoi elektricheskoi promyshlennosti', *Elektrichestvo*, 1900, 15–16: 202.

9. Holzer, op. cit. (3); 'Vyvoz iz Germanii produktov elektrotekhnicheskoi promyshlennosti', *Elektrichestvo*, 1914, 13: 347–50.

10. Walther Kirchner, 'Russian tariffs and foreign enterprises before 1919: the German entrepreneur's perspective', *Journal of European History*, 1981, 11: 361–80.

11. A. Lomakin, 'Privoz v Russiiu produktov elektrotekhnicheskoi promyshlennosti', *Elektrichestvo*, 1912, 10: 300.

12. Diakin, op. cit. (3), 25, 84; Gusev, op. cit. (3), 12; Antony C. Sutton, *Western Technology and Soviet Economic Development, 1917 to 1930*, (Stanford, 1968), 223.

13. For enemy countries, 'Vyvoz iz Astro-Vengerii produktov elektricheskogo promyshlennosti v 1913 gody', *Elektrichestvo* 1914, 13: 336–7; and 'Vyvoz iz Germanii produktov elektricheskoi promyshlennosti v 1913 gody', *Elektrichestvo*, 1914, 14: 347–9; also 'Obshchii ocherk zadach, organizatsii i deiatelnosti elektrotekhnicheskogo otdela Tsentralnogo Voenno-Promyshlennogo Komiteta v sviazi s polozheniem elektrotekhnicheskoi promyshlennosti v Rossii do voiny', *Elektrichestvo*, 1916, 3: 34.

14. Walther Kirchner, 'The industrialization of Russia and the Siemens firm, 1853–1890', *Jahrbucher für Geschichte Osteuropas*, 1974, 22: 321–57.

15. Central State Historical Archives of Leningrad (TsGIAL), f. 20, o. 4, ed.kh. 3594, 1, 73.

16. 'Dannye o razvitie Obshchestva elektricheskogo osveshcheniia 1886 goda', *Elektrichestvo*, 1915, 5: 99–100.

17. S.A. Gusev, 'Pervaia promyshlennaia ustanovka trekhfaznogo toka v Rossii', *Trudy po istorii tekhniki*, 1953, 6: 74–84.

18. M.O. Kamenetskii, *Robert Eduardovich Klasson*, (Moscow, 1963), 22, 25–32, 50–4.

19. Central State Archives of the National Economy (TsGANKh), f. 9508, o. 1, ed.kh. 14, 4.

20. T.F. Makarev, 'Razvitie oborudovaniia tsentralnykh elektricheskikh stantsii v Peterburge', *Elektrichestvo*, 1912, 6: 181.

21. 'Statisticheskie svedeniia o tsentralnykh elektricheskikh stantsii v Rossii za 1913 god', *Elektrichestvo,* 1915, 11: 206–15.

22. For the tests and first use, M.D. Kamenetskii, *Pervye Russkie elektrostantsii*, (Leningrad, 1951), 50. For the development of specialized coal grates, 'Razvitie oborudovaniia tsentralnykh elektricheskikh stantsii v Peterburge', *Elektrichestvo*, 1912, 6: 188.

23. E.R. Ulman, 'Razvitie tsentralnykh elektricheskikh stantsii v Peterburge za desiatiletnyi period', *Elektrichestvo*, 1912, 4: 118.

24. V.I.K. 'Belgiiskie kapitaly v Rossii', *Zapiski Imperialskogo Russkogo Tekhnicheskogo Obshchestvo*, 1909, 5: 154.

25. P.V. Barabanov, 'Pervyi elektricheskii tramvai v Rossii', *Pochtovo-Telegraficheskii zhurnal*, 1893, 5: 466–82.

26. V. Shelgunov, 'Belgiiskii tramvai v Rossii', *Izvestiia Moskovskoi gorodskoi dumy*, 1914, 3: 71.

27. Ibid., 81.

28. Diakin, op. cit. (3), 268–9.

29. Arcadius Kahan, 'Capital formation during the period of early industrialization in Russia, 1890–1913', *Cambridge Economic History of Europe*, (Cambridge, 1978), VII, 2: 273; P.V. Ol, *Foreign Capital in Russia*, Geoffrey Jones and Grigori Gerenstain, trans., (New York, 1983), 9; Fred V. Cartensen, 'Numbers and reality: a critique of foreign investment estimates in Tsarist Russia', in Maurice

Levy-Leboyer (ed.), *La Position Internationale de la France*, (Paris, 1977), 275–83; John P. McKay, 'Foreign enterprise in Russian and Soviet industrialization: a long-term perspective', *Business History Review*, 1974, 48: 336–56.

30. Diakin, op. cit. (3), 41–4.

31. Ibid., 268–9. Diakin excluded an unknown number of municipal operations and domestic concessions, thereby somewhat understating the Russian contribution.

32. Ibid.

33. *Potrebitelskie Elektricheskie Stantsii*, (Moscow, 1913), 3; Alfred J. Rieber, *Merchants and Entrepreneurs in Imperial Russia*, (Chapel Hill, 1982), 105.

34. M. Giterman, 'Elektrichestvo i munitsipalitety', *Izvestiia Moskovskoi gorodskoi dumy*, 1914, 11: 64; Christopher Armstrong and H.V. Nelles, *Monopoly's Moment: The Organization and Regulation of Canadian Utilities, 1830–1930*, (Philadelphia, 1986), 116; Thomas P. Hughes, 'The electrification of America: the system builders', *Technology and Culture*, 1979, 28: 155; A.J. Millard, *A Technological Lag: Diffusion of Electrical Technology in England, 1879–1914*, (New York, 1987), 155–6.

35. Clive Trebilcock, *The Industrialization of the Continental Powers, 1780–1914*, (London, 1981), 268, 290.

36. Data compiled from *Elektrichestvo* obituaries and *Great Soviet Encyclopedia* articles.

37. Belkind, op. cit. (8); 'P.N. Jablochkov: Nekrolog', *Elektrotekhnicheskii Vestnik*, 1894, 4: 121–2; 'The Jablochkoff system of electric illumination', *Engineering*, 26 July 1878, 63–5.

38. Oleg N. Veselovskii, *Dolivo-Dobrovolskii, 1862–1919*, (Moscow, 1963); 'Nekrolog', *Elektrichestvo*, 1930, 5: 258–9.

39. 'Achilles de Khotinsky', *National Cyclopaedia of American Biography*, (New York, 1936), XXV, 63–4; A. Heerding, *The History of N.V. Philips' Gloeilampenfabrieken: The Origin of the Dutch Incandescent Lamp Industry*, vol. 1, (Cambridge, 1986), 139–40, 148.

40. M.A. Shatelen, 'Iz "Ankety sredi inzhener-elektrikov" okonchivshikh STP Politekhnicheskii Institut Imperatora Petra Velikogo', *Elektrichestvo*, 1914, 4: 130.

41. Russians used the German *schwachstrom* ('weak current') and *starkstrom* ('strong current') to distinguish between telecommunications (telegraph and telephone) and the power industry.

42. Ibid., 136.

43. 'Statisticheskie svedeniia o tsentralnykh elektricheskikh stantsii v Rossii', *Elektrichestvo*, 1910, 1: 1.

44. N.N. Georgievskii, 'Normalizatsiia i standartizatsiia v oblasti elektrotekhniki v SSSR', *Elektrichestvo*, 1925, 1: 24–8; N.A. Shostin, 'Iz istorii elektricheskikh etalonov', *Elektrichestvo*, 1945, 7: 6–9.

45. 'Deiatelnost VI', *Elektrichestvo*, 1916, 7–8: 141.

46. 'Khronika', *Elektrichestvo*, 1917, 9–10: 145.

47. 'Raznye izvestiia', *Elektrichestvo*, 1888, 17–19: 176. The destruction of VDE membership records in the Second World War makes full knowledge of Russian involvement impossible (personal communication, 26 June 1982).

48. For example, Russian membership in the British Institution of Electrical Engineers varied from none to three from 1872 to 1915. Data produced by Geoffrey Tweedale for W.J. Reader's *A History of the Institution of Electrical Engineers*, (London, 1987).

49. 'Po russkim gorodam', *Elektrichestvo*, 1917, 1: 26.

50. TsGIAL f. 23, o. 27, d. 70, 34–35.

51. G.A. Feldman, 'Elektrobank', *Voprosy Elektrifikatsii*, 1922, 1–2: 66–71.

52. M. Giterman, 'Elektrichestvo i munitsipalitety', *Izvestiia Moskovskoi gorodskoi dumy*, 1914, 8: 71.

53. Rieber, op. cit. (33), 355.

54. For example, I.Ia. Perelman, *Elektricheskaia energiia i melkoe proizvodstvo*, (Moscow, 1906), 3-22; M. Giterman, 'Elektrichestvo i munitsipalitety', *Izvestiia Moskovskoi gorodskoi dumy*, 1914, 8: 52-71, 9: 63-78, and 10: 54-72; O.G. Flekkel, ' "Ekonomischeskie" lampochki i populiarizatsiia elektrichestva', *Gorodskoe Delo* 1912, 4: 237; V. Smirnov, 'Novaia forma eksploatatsii gorodskikh predpriiatii: ni kontsessiia, ni munitsipalizatsiia', in 'Deiatelnost uchenykh i tekhnicheskikh obshchestv', *Elektrichestvo*, 1914, 10: 252-3.

55. Armstrong and Nelles, *op. cit.* (34), 141-58; Anthony Sutcliffe, *Toward the Planned City: Germany, Britain, the United States and France, 1780-1914*, (New York, 1981).

56. G.M. Krzhizhanovskii, 'Oblastnye elektricheskie stantsii na torfe i ikh znachenie dlia tsentralnogo promyshlennogo raiona', *Izbrannoe*, (Moscow, 1957), 16.

57. V.I. Grinevetskii, *Tekhniko-Obshchestvennye zadachi v sfere promyshlennosti i tekhniki v sviazi s voinoi*, (Moscow, 1914), 13-15.

58. V.I. Kovalevskii, 'Osnovnye nuzhdy russkoi promyshlennosti', *Trudy komissii po promyshlennosti v sviazi s voinoi*, 1915, 5: 7-8.

59. Ruth A. Roosa, 'Russian industrialists and "state socialism", 1906-17', *Soviet Studies*, 1972, 23: 414-16; Kendall E. Bailes, *Technology and Society under Lenin and Stalin: Origins of the Soviet Technical Intelligentsia, 1917-1941*, (Princeton, 1978), 22-3, 424.

60. Richard Stites, *Revolutionary Dreams: Utopian Vision and Experimental Life in the Russian Revolution*, (New York, 1989), 36-7, 45.

61. 'Khronika', *Narodnoe Khoziaistvo*, 1918, 2: 19.

62. Ibid.; also 'Deiatelnost elektrotekhnicheskogo otdela V.S.N.Kh.', *Narodnoe Khoziaistvo*, 1919, 1-2: 41-2.

63. 'Khronika', *Narodnoe Khoziaistvo*, 1918, 2: 19. Emphasis added.

64. Peter Rutland, *The Myth of the Plan: Lessons of Soviet Planning Experience*, (London, 1985), 11; Leon Smolinski, 'Planning without theory', *Survey*, 1967, 64: 116; Alek G. Cummins, 'The road to NEP, the State Commission for the Electrification of Russia (GOELRO): a study in technology, mobilization and economic planning', (unpublished PhD thesis, University of Maryland, 1988, AA08818383), 52-3.

65. Vladimir I. Lenin, *Collected Works*, 4th edn, (Moscow, 1966), 32: 140; Carr, *The Bolshevik Revolution*, vol. 2, 373; Smolinski, 'Planning without theory', 117-20; Roger W. Pethybridge, *The Social Prelude to Stalinism*, (London, 1974), 43; 'Karl Ballod', *Bolshaia Sovietskaia Entsiklopediia*, 1926, 4: 539-40.

66. Aleksandr I. Ugrimov, 'Moi put i rabota v GOELRO', *Sdelaem Rossiiu elektricheskoi*, (Moscow, 1961), 85.

67. G.M. Krzhizhanovskii, 'Tekushchie voprosy elektrifikatsii', *Elektrichestvo*, 1922, 2: 4; also L. Dreier, *Zadachi i Razvitie Elektrotekhniki*, (Moscow, 1919), 22, and Boris Kushner, *Revoliutsiia i Elektrifikatsiia*, (Petrograd, 1920), 13.

68. G. Klingenberg, 'Electricity Supply in Large Cities', *The Electrician*, 12 December 1913, 398-401; G. Klingenberg, *Large Electric Power Stations: Their Design and Construction*, (London, 1916); 'Lignite as station fuel', *Electrical Review*, 14 March 1913, 419 and 30 May 1913; Edmund N. Todd, 'Industry, state, and electrical technology in the Ruhr *circa* 1900', *Osiris*, 1989, 5: 243-59.

69. *Vosmoi vserossiiskii sezd sovietov: Stenograficheskii otchet*, (Moscow, 1921), 30.

70. 'Velikii Ekonomicheskii Plan', in *Plan Elektrifikatsii RSFSR. Doklad VIII Sezdi Sovietov Gosudarstvennoi Komissii po Elektrifikatsii Rossii*, 2nd edn, (Moscow, 1955), 16.

71. *Trudy GOELRO. Dokumenty i materialy,* (Moscow, 1960), 26 October, 179; see also 'Inostrannyi kapital i elektrifikatsiia Rossii', *Trud*, 15 February 1922, 4.
72. 'Protokol', 24 April, *Trudy GOELRO*, 128-9.
73. Ibid., 22 May, 179.
74. Ibid.
75. *50 Let leninskogo plana GOELRO. Sbornik materialov*, (Moscow, 1970), 89-90.
76. For postwar efforts of the Academy of Science to restore foreign links, see Elizaveta D. Lebedkina, *Mezhunarodnyi Soviet Nauchnykh Soiuzov i Akademii Nauk SSSR*, (Moscow, 1974), 35-57.
77. *Elektrifikatsiia Rossii. Trudy 8 Vserossiiskogo Elektrotekhnicheskogo Sezda*, (Moscow, 1921), I, 5. This statement was part of the official introduction.
78. Ibid., Krug's speech on the electrotechnical industry, I, 69.
79. Ibid.
80. Ibid., I, 28.
81. Ibid., II, 5, introduction to the technical-economic section.
82. Ibid., Krzhizhanovskii, I, 89; Kozmin, I, 79.
83. Ibid., M.A. Shatelen and B.E. Borovev, 'Snabzhenie selsko-khoziaistvennykh raionov elektricheskoi energii ot mestnykh stantsii maloi moshchnosti', II, 53-4.
84. A. Brauner, 'K voprosu o peredache elektricheskikh stantsii obshchogo polzovanii elektrootdelam', *Kommunalnoe Delo*, 1921, 1: 37-8.
85. *Elektrifikatsiia Rossii*, op. cit. (81), F.K. Ryndin, 'O vozmozhnosti ustroistva elektricheskogo osveshchaniia v derevne v sviazi i sozdaniem semi raionnykh melnits', II, 24-34.
86. Ibid.; Kozmin, I, 79.
87. Ibid.; E. Liskun, 'Selskoe khoziaistvo Severnoi oblasti v sviazi v planami elektrifikatsii', II, 72-86.
88. Ibid., I, 163.
89. Kozmin, 'K itogam VIII Vseross. Elektrotekhn. Sezda', *Narodnoe Khoziaistvo*, 1921, 11-12: 105. The problem with this analogy is that the American North's industrial strength remained untouched by the horrors of war; the agrarian South, on the other hand, suffered greatly.
90. Charles S. Maier, 'Between Taylorism and technocracy: European ideologies and the vision of industrial productivity in the 1920s', *Journal of Contemporary History*, 1970, 5: 27-61; William E. Akin, *Technocracy and the American Dream: the Technocratic Movement, 1900-1941*, (Berkeley, 1977), 4, 46.
91. For the Ruhr, Edmund N. Todd, 'Technology and interest group politics: electrification of the Ruhr, 1886-1930' (unpublished PhD thesis, University of Pennsylvania, 1984, AA08422040). For Giant Power, Thomas P. Hughes, 'Technology and public policy', *IEEE Proceedings*, 1976, 64: 1361-71, and Bayla Singer, 'Power politics', *IEEE Technology and Society Magazine*, December 1988, 20-7. For Superpower, Terry Kay Rockefeller, 'The failure of planning for electrical power supply: the case of the electrical engineers and "superpower," 1915-1924', in Joel A. Tarr (ed.), *Retrospective Technology Assessment—1976*, (San Francisco, 1977), 191-215.
92. Thomas Remington, *Building Socialism in Bolshevik Russia: Ideology and Industrial Organization, 1917-1921,* (Pittsburgh, 1984), 19, 117-18, 136-39; Stites, op. cit. (63), 52, 145.
93. Rainer Traub, 'Lenin and Taylorism: the fate of "scientific management" in the (early) Soviet Union', *Telos*, 1978, 37: 87.
94. E.Ia. Shulgin, 'K peresmotru plana elektrifikatsii', *Planovoe khoziaistvo*, 1925, 2: 22-3.
95. 'Mestnaia elektrifikatsiia', *Elektrifikatsiia*, 1924, 9-10: 37; Elektroselstroi,

Elektroselstroi i ego deiatelnost, (Moscow, 1924), 7-8.

96. E.H. Carr, *The Bolshevik Revolution, 1917-1923*, (London, 1951), vol. 1, 276-89; Maurice Dobb, *Soviet Economic Development since 1917*, (New York, 1948), 150-1.

97. Charles Bettelheim, *Class Struggles in the USSR. The Second Period: 1923-1930*, trans. by Brian Pearce, (New York, 1978), 58-9.

98. A.Z. Goltsman and A.A. Gorev, 'Plan elektrifikatsii i krestianskoe khoziaistvo', *Planovoe Khoziaistvo*, 1925, 4: 178.

99. 'Iz zhizni', *Elektrichestvo*, 1922, 1: 51.

100. A. Kulikovskii, 'O role gosudarstva v elektrifikatsii derevni', *Ekonomicheskaia Zhizn*, 22 April 1922, 3.

101. Elektrobank, *Finansirovanie elektrokhoziaistva (dva goda raboty Elektrobanka)*, (Moscow-Leningrad, 1927), 6.

102. Glavelektro, *Obzor sostoianiia rabot po krupnomu elektrostroitelstvu na 1oe Oktiabria 1925 g*, (Moscow, 1926), 19, 22, 24, 26, 29, 30, 33.

103. For example, Tomsk received a foreign turbo-generator more quickly at half the cost of a Soviet system ('Kommunalnye predpriiatiia', *Kommunalnoe Delo*, 1925, 20: 53).

104. M. Tipograf, 'Elektrotekhnicheskaia promyshlennosti', in A.M. Ginzberg (ed.), *Chastnyi kapital v narodnom khoziaistve SSSR*, (Moscow, 1927), 548.

105. P.I. Voevodin, 'Na Volkhovstroe', *Elektrifikatsiia*, 1924, 2: 24.

106. Hans Rogger, '*Amerikanizm* and the economic development of Russia', *Journal for the Comparative Study of Society and History*, 1981, 23: 388-9; Traub, op. cit. (93), 82-92; Thomas P. Hughes, *American Genesis: A Century of Invention and Technological Enthusiasm*, (New York, 1989), 249-84.

107. Kendall E. Bailes, 'The American connection: ideology and the transfer of American technology to the Soviet Union, 1917-1941', *Journal for the Comparative Study of Society and History*, 1981, 23: 429.

108. Alexander Elrich, *The Soviet Industrialization Debate, 1924-1928*, (Cambridge, 1960); Moshe Lewin, *Political Undercurrents in Soviet Economic Debates*, (Princeton, 1974); Keith Smith, 'Economic theory and the closure of the Soviet industrialization debate', in Keith Smith (ed.), *Soviet Industrialization and Soviet Maturity*, (London, 1986), 23-49; E.H. Carr and R.W. Davies, *Foundations of a Planned Economy, 1926-1929*, vol. 1, (New York, 1969), 271-4.

109. Carr, *Socialism in One Country, 1924-1926*, vol. 1, 508; R.W. Davies, *The Soviet Economy in Turmoil, 1929-1930*, (Cambridge, MA, 1989), 47-9.

110. Gleb M. Krzhizhanovskii, 'Perspektivy elektrifikatsii', in Gleb M. Krzhizhanovskii, A.A. Gorev and V.Z. Esin, *Chetyre goda elektrifikatsii SSSR*, (Moscow, 1925), 15, 21.

111. Elektrobank, *Nekotorye itogi deiatelnosti Elektrobanka*, (Moscow, 1925), 5.

112. R.W. Davies, *The Soviet Economy in Turmoil, 1929-1930*, (Cambridge, 1989), 47-8.

ICOHTEC XVIII Conference Report

A Personal View

The Conservatoire des Arts et Métiers, Paris, was the majestic setting for the ICOHTEC XVIII conference of 8–14 July 1990, attended by something approaching 150 participants from seventeen countries. The first two days of the formal programme were devoted to two symposia: on 'Gustave Adolphe Hirn' and 'Engineering and the shaping of the natural and urban environment', with the remaining three days of the programme taken up with 'General themes', or, in other words, topics too diverse to be easily classifiable, if indeed classifiable at all. The action 'off-stage', the renewing of old friendships and the making of new ones, was, as always, at least as important a part of the proceedings as its more visible public parts (the next opportunity for interaction, whether public or private, will be, it should be noted, in Vienna in September 1991).

Twenty-nine papers were offered under the rubric 'General themes'. Plainly they would be rather too large a number to review serially even if the present reporter felt omni-competent enough to attempt it. Instead I shall take the easier (but inescapably invidious) course of selecting just two papers that seem to me to have touched on one of the big issues of ecumenical importance in the history of technology. This is the important question concerning the driving force or (altogether more likely) the multiplicity of factors, acting synergistically, that propel first one culture or people and then another (over considerable time scales) out of the ruck and into a position of technological primacy. After every parade, however, follows dustcart and broom. Only second in importance to this area of enquiry, therefore, comes what one might call post-mortem studies: the search to understand how it is that technologically hegemonic areas or 'empires' lose their dynamic, with consequences ranging from mere downturn into relative decline to that of full-blown Ozymandian-style eclipse.

As to the first theme it seemed to me that the paper by Dr Yukitoshi Matsuo of Kyoto, Japan, on 'The making of "science–technology" faith in Japan', had (potentially) things to relate of exceptional importance regarding the relationship of technology and culture of Japan. Everyone knows well enough that the two are related but who would claim that our understanding of the relationship, *the* macro-question in the history of technology, is anything but rudimentary? All the more regrettable, therefore, that Dr Matsuo, supplying no printed abstract of his paper, should also have spoken in so opaque an English as to cloud the manifest importance of what he had to say. The result was like a classical Chinese landscape painting: fleeting glimpses of some pretty exciting scenery seen through a wrack of mists and exhalations.

If, however, I understood Dr Matsuo sufficiently well, then although the word 'faith' does not satisfactorily translate the sense of the Japanese word that he was trying to render, nor 'religion' either for that matter, the reception of Western science and technology by the Japanese after 1868 was certainly akin to a religious experience, or perhaps, one should say, a revelation, because the new 'nature of things' that was then disclosed was perceived as an amplification of the Confucian tradition. Whereas the Emperor had been regarded as the embodiment of the universal source of order, by degrees, but with gathering pace, science and technology was to displace the emperor cult. In a word, because science and technology is conformable to the moral and ethical tradition of Japan, striving for the best possible quality in industrial manufactures is in spiritual terms nothing less than a matter of faith and as such allows no duality to appear between what one is and what one makes. There is, however, a group aspect to all this: the individual is not simply confronting his conscience in isolation, as would be the case in the West where knowledge of what exact craftsmanship (loosely understood) dictates is held *in petto* as a guiding force. For the Japanese, poor-quality workmanship amounts to an unacceptable social act, striking at one's social esteem much more severely than, say, a breach of good manners such as spitting in public would be. It follows from this that the Western monastic ideal of labour as prayer, the Benedictine *laborare est orare*, provides only a partial insight into the Japanese mode of being. If Dr Matsuo is correct (and also, that is, if I have followed him sufficiently well) then Caucasians and others might as well throw up their hands in dismay (workers of the world, despair!) as hope to match Japanese industrial performance. How well or ill such 'faith' may equate with green issues is another question, one not considered by Dr Matsuo, let alone answered. It would be fascinating indeed to know whether enquiries of this nature have been or are being undertaken.

On the question of technological decline the paper by Dr David Edgerton of Manchester on 'British industrial research and development 1900–1970' offered some insight into the reasons for Britain's secular decline from a position of technological hegemony. I hope my selection of this paper will not be taken as parochial. In fact, in broad historiographical terms it performs a useful function as counterpoint to the paper on Japan. If 'faith' may be held to explain the rise of Japan, might lack of 'faith' be involved in Britain's decline? Dr Edgerton would perhaps answer that if faith in science and technology is manifested, at least in part, in R&D expenditures, then in this respect at least there was no lack of faith in Britain in the period under scrutiny. He concluded in fact that after 1945, 'British businesses probably overinvested in research and development'. This was a valuable contribution to the debate even if its conclusions were negative.

One object of making the life and work of Gustave-Adolphe Hirn (1815–90) the subject of a symposium was to honour him by marking the centenary of his death. Another was to draw attention to the fact that no biography had yet done justice to the part he played in nineteenth-century technological development. In opening the symposium Dr Jacques Payen of Paris expressed the hope that the papers to be presented would stimulate further research into Hirn's contribution to science and industry. A logical mode of enquiry would require the scrutinizing in sequence of, first, his intellectual formation as a member of a family of industrial entrepreneurs; then his own contribution to the textile industry and to his pioneering work in developing lubricants based on mineral oil. Paralleling these interests attention would need to be paid to his more purely scientific researches into friction, the mechanical equivalent of heat, and into the thermal properties of steam engines. Last, but not least, after his retirement from active industrial involvement in 1880, would come his important experiments on the physical properties of gases and his development (strange hybrid) of a '*metaphysique expérimentale*'. The papers following Dr Payen's fell into two groups. The first dealt with Hirn's scientific work, with his lectures of 1864 on the theory of heat, his *L'ésquisse élémentaire de la théorie mécanique de la chaleur*, his development of calorimetry for industrial purposes, and the sources of his *L'analyse élémentaire de l'univers* of 1868. A second series dealt with the development of Hirn's work on locomotives, on thermodynamic performance in relation to the walls of cylinders and on double expansion as applied to marine engines and locomotives. What was missing here, to my surprise, was any mention of his '*transmission télé-dynamique*', the transmission of power, sometimes over thousands of metres, by means of iron and (later) steel cables running on specially designed pulleys. Like earlier systems of power transmission its fate was sealed by the advent of electrical power but it certainly created a stir in the technological literature of the day and was of considerable industrial importance. Hirn's papers on these and other matters are housed in the library of the Société industrielle de Mulhouse, which society's own journal (not readily to be consulted in England) forms one of the essential sources for the history of nineteenth-century European technology.

The second of the two symposia, on 'Le rôle des ingénieurs dans l'aménagement des territoires et des villes', was introduced by Dr André Guillerme. His theme was the international influence exerted by the great French engineering schools, the *École des Ponts et Chaussées* and the *École militaire*, in the establishment of similar institutions abroad. In the USA West Point was modelled on the latter; the former served as model for the St Petersburg Institute of Highway Engineering in Tsarist Russia, to whose staff young French engineers were recruited with the rank of Lieutenant-Colonel. Prussia followed with its Bau-Akademie in Berlin in 1824. Other papers dealt with a wide variety of topics falling within the rubric; among these were three pieces dealing with problems of town water supply. The paper of Dr Hugh Torrens of Keele on John Rennie's stone water pipes was at once learned, amusing and a useful addition to the gallery of failed innovations. For those whose interest is piqued by the notion of failed innovations and who sense its heuristic potential I should mention here that the symposium papers on this subject held in Hamburg in 1989 as part of the International Congress of the History of Science will probably, by the time this report appears, have been edited by Professor Hans Braun and published in *Social Studies of Science*.

Graham Hollister-Short

Contents of Former Volumes

FIRST ANNUAL VOLUME, 1976

D.S.L. CARDWELL and RICHARD L. HILLS, Thermodynamics and Practical Engineering in the Nineteenth Century.

JACQUES HEYMAN, Couplet's Engineering Memoirs, 1726–33.

NORMAN A.F. SMITH, Attitudes to Roman Engineering and the Question of the Inverted Siphon.

R.A. BUCHANAN, The Promethean Revolution: Science, Technology and History.

M. DAUMAS, The History of Technology: its Aims, its Limits, its Methods.

KEITH DAWSON, Electromagnetic Telegraphy: Early Ideas, Proposals and Apparatus.

MARIE BOAS HALL, The Strange Case of Aluminium.

G. HOLLISTER-SHORT, Leads and Lags in Late Seventeenth-century English Technology.

SECOND ANNUAL VOLUME, 1977

EMORY L. KEMP, Samuel Brown: Britain's Pioneer Suspension Bridge Builder.

DONALD R. HILL, The Banū Mūsà and their 'Book of Ingenious Devices'.

J.F. CAVE, A Note on Roman Metal Turning.

J.A. GARCIA-DIEGO, Old Dams in Extremadura.

G. HOLLISTER-SHORT, The Vocabulary of Technology.

RICHARD L. HILLS, Museums, History and Working Machines.

DENIS SMITH, The Use of Models in Nineteenth-century British Suspension Bridge Design.

NORMAN A.F. SMITH, The Origins of the Water Turbine and the Invention of its Name.

THIRD ANNUAL VOLUME, 1978

JACK SIMMONS, Technology in History.

R.A. BUCHANAN, History of Technology in the Teaching of History.

P.B. MORICE, The Role of History in a Civil Engineering Course.

JOYCE BROWN, Sir Proby Cautley (1802–71), a Pioneer of Indian Irrigation.

A. RUPERT HALL, On knowing, and knowing how to . . .

FRANK D. PRAGER, Vitruvius and the Elevated Aqueducts.

JAMES A. RUFFNER, Two Problems in Fuel Technology.

JOHN C. SCOTT, The Historical Development of Theories of Wave-Calming using Oil.

FOURTH ANNUAL VOLUME, 1979

P.S. BARDELL, Some Aspects of the History of Journal Bearings and their Lubrication.

K.R. FAIRCLOUGH, The Waltham Pound Lock.

ROBERT FRIEDEL, Parkesine and Celluloid: The Failure and Success of the First Modern Plastic.

J.G. JAMES, Iron Arched Bridge Designs in Pre-Revolutionary France.

L.J. JONES, The Early History of Mechanical Harvesting.

G. HOLLISTER-SHORT, The Sector and Chain: An Historical Enquiry.

FIFTH ANNUAL VOLUME, 1980

THOMAS P. HUGHES, The Order of the Technological World.

THORKILD SCHIØLER, Bronze Roman Pistol Pumps.

STILLMAN DRAKE, Measurement in Galileo's Science.

L.J. JONES, John Ridley and the South Australian 'Stripper'.

D.G. TUCKER, Emile Lamm's Self-Propelled Tramcars 1870–72 and the Evolution of the Fireless Locomotive.

S.R. BROADBRIDGE, British Industry in 1767: Extracts from a Travel Journal of Joseph Banks.

RICHARD L. HILLS, Water, Stampers and Paper in the Auvergne: A Medieval Tradition.

SIXTH ANNUAL VOLUME, 1981

MARJORIE NICE BOYER, Moving Ahead with the Fifteenth Century: New Ideas in Bridge Construction at Orleans.

ANDRÉ WEGENER SLEESWYK, Hand-Cranking in Egyptian Antiquity.

CHARLES SÜSSKIND, The Invention of Computed Tomography.

RICHARD L. HILLS, Early Locomotive Building near Manchester.

L.L. COATSWORTH, B.I. KRONBERG and M.C. USSELMAN, The Artefact as Historical Document. Part 1: The Fine Platinum Wires of W.H. Wollaston.

A. RUPERT HALL and N.C. RUSSELL, What about the Fulling-Mill?

MICHAEL FORES, *Technik*: Or Mumford Reconsidered.

SEVENTH ANNUAL VOLUME, 1982

MARJORIE NICE BOYER, Water Mills: a Problem for the Bridges and Boats of Medieval France.

Wm. DAVID COMPTON, Internal-combustion Engines and their Fuel: a Preliminary Exploration of Technological Interplay.

F.T. EVANS, Wood Since the Industrial Revolution: a Strategic Retreat?

MICHAEL FORES, Francis Bacon and the Myth of Industrial Science.

D.G. TUCKER, The Purpose and Principles of Research in an Electrical Manufacturing Business of Moderate Size, as Stated by J.A. Crabtree in 1930.

ROMAN MALINOWSKI, Ancient Mortars and Concretes: Aspects of their Durability.

V. FOLEY, W. SOEDEL, J. TURNER and B. WILHOITE, The Origin of Gearing.

EIGHTH ANNUAL VOLUME, 1983

W. ADDIS, A New Approach to the History of Structural Engineering.

HANS-JOACHIM BRAUN, The National Association of German-American Technologists and Technology Transfer between Germany and the United States, 1884–1930.

W. BERNARD CARLSON, Edison in the Mountains: the Magnetic Ore Separation Venture, 1879–1900.

THOMAS DAY, Samuel Brown: His Influence on the Design of Suspension Bridges.

ROBERT H.J. SELLIN, The Large Roman Water Mill at Barbegal (France).

G. HOLLISTER-SHORT, The Use of Gunpowder in Mining: A Document of 1627.

MIKULÁŠ TEICH, Fermentation Theory and Practice: the Beginnings of Pure Yeast Cultivation and English Brewing, 1883–1913.

GEORGE TIMMONS, Education and Technology in the Industrial Revolution.

NINTH ANNUAL VOLUME, 1984

P.S. BARDELL, The Origins of Alloy Steels.

MARJORIE NICE BOYER, A Fourteenth-Century Pile Driver: the *Engin* of the Bridge at Orleans.

MICHAEL DUFFY, Rail Stresses, Impact Loading and Steam Locomotive Design.

JOSÉ A. GARCIA-DIEGO, Giovanni Francesco Sitoni, an Hydraulic Engineer of the Renaissance.

DONALD R. HILL, Information on Engineering in the Works of Muslim Geographers.

ROBERT J. SPAIN, The Second-Century Romano-British Watermill at Ickham, Kent.

IAN R. WINSHIP, The Gas Engine in British Agriculture, *c.* 1870–1925.

TENTH ANNUAL VOLUME, 1985

D. de COGAN, Dr E.O.W. Whitehouse and the 1858 trans-Atlantic Cable.

A. RUPERT HALL, Isaac Newton's Steamer.

G.J. HOLLISTER-SHORT, Gunpowder and Mining in Sixteenth- and Seventeenth-Century Europe.

C.J. JACKSON, Evidence of American Influence on the Designs of Nineteenth-Century Drilling Tools, Obtained from British Patent Specifications and Other Sources.

JACQUES PAYEN, Beau de Rochas Devant la Technique et l'Industrie de son Temps.

ÖRJAN WIKANDER, Archaeological Evidence for Early Water-Mills—an Interim Report.

A.P. WOOLRICH, John Farey and the Smeaton Manuscripts.

MIKE CHRIMES, Bridges: a Bibliography of Articles Published in Scientific Periodicals 1800–1829.

ELEVENTH ANNUAL VOLUME, 1986

HANS-JOACHIM BRAUN, Technology Transfer Under Conditions of War: German Aero-technology in Japan During the Second World War.

VERNARD FOLEY, with SUSAN CANGANELLI, JOHN CONNOR and DAVID RADER, Using the Early Slide-rest.

J.G. JAMES, The Origins and Worldwide Spread of Warren-truss Bridges in the Mid-nineteenth Century. Part 1: Origins and Early Examples in the UK.

ANDREW NAHUM, The Rotary Aero Engine.

DALE H. PORTER, An Historian's Judgments About the Thames Embankment.

JOHN H. WHITE, More Than an Idea Whose Time Has Come: The Beginnings of Steel Freight Cars.

IAN R. WINSHIP, The Acceptance of Continuous Brakes on Railways in Britain.

TWELFTH ANNUAL VOLUME, 1990

KENNETH C. BARRACLOUGH, Swedish Iron and Sheffield Steel.

IAN INKSTER, Intellectual Dependency and the Sources of Invention: Britain and the Australian Technological System in the Nineteenth Century.

M.T. WRIGHT, Rational and Irrational Reconstruction: The London Sundial-Calendar and the Early History of Geared Mechanisms.

J.V. FIELD, Some Roman and Byzantine Portable Sundials and the London Sundial-Calendar.

R.T. McCUTCHEON, Modern Construction Technology in Low-income Housing Policy: The Case of Industrialized Building and the Manifold Links between Technology and Society in an Established Industry.

Book Review by Frank A.J.L. James: André Guillerme, *Le Temps de l'Eau: La Cité, l'Eau et les Techniques: Nord de la France Fin IIIe–Début XIXe Siècle*. Eng. trans.: *The Age of Water: The Urban Environment in the North of France, AD 300–1800.*